Brazilian Environmental and Climate Change Law

Rafael Martins Costa Moreira • Gabriel Wedy

Brazilian Environmental and Climate Change Law

Springer

Rafael Martins Costa Moreira
Justiça Federal do Rio Grande do Sul
Porto Alegre, Rio Grande do Sul, Brazil

Gabriel Wedy
Law, UNISINOS - Universidade do Vale do Rio dos Sinos
Porto Alegre, Rio Grande do Sul, Brazil

ISBN 978-3-031-87997-5 ISBN 978-3-031-87998-2 (eBook)
https://doi.org/10.1007/978-3-031-87998-2

This Springer imprint is published by the registered company Springer Nature Switzerland AG
The registered company address is: Gewerbestrasse 11, 6330 Cham, Switzerland

Foreword

In this groundbreaking book, Federal Judges and Professors Gabriel Wedy and Rafael Martins Costa Moreira present and translate the history, state-of-the-art, and future prospects of Brazilian environmental and climate law into the English language. Designed particularly for an international audience, this book offers readers an in-depth exploration of how Brazilian law governs the relationship between humanity and the environment, emphasizing that development must necessarily align with the ecological sustainability mandated by the Federal Constitution (Art. 170, VI). Legal professionals, especially judges, must remain vigilant in applying the constitutional provisions set forth in Article 225 of the Citizen Constitution, as well as the corresponding constitutional duties to safeguard the environment and the climate system. Equally critical is the enforcement of the extensive and protective infra-constitutional environmental legislation, which requires judges to adopt not only an intragenerational perspective but also an intergenerational one, ensuring the preservation of our Earth—our *Gaia*—for future generations.

Through this book, readers will delve into the intricacies and challenges of Brazilian environmental policy, which confronts greenhouse gas emissions and diverse forms of pollution. Beyond its continental dimensions and multicultural, multiracial population, Brazil is home to the world's largest tropical rainforest, the Amazon, with its unparalleled and rich biodiversity. The backdrop of this book is a nation teeming with diverse flora and fauna and a vibrant plurality of indigenous and traditional communities with profound cultural richness. Brazil holds immense potential to significantly reduce global greenhouse gas emissions, particularly through carbon sequestration enabled by preserving native vegetation, combating deforestation, and preventing wildfires. The nation also boasts a significant share of the world's freshwater resources and the planet's greatest biodiversity, while simultaneously serving as a global leader in food production, a crucial element in addressing global hunger.

Brazil's strategic environmental leadership, however, would be gravely jeopardized without a robust legal and regulatory framework to protect its ecosystems. As often observed, Brazil possesses some of the most advanced and comprehensive environmental legislation in the world. Yet, without proper implementation, these laws are rendered meaningless.

Brazilian environmental legislation and jurisprudence highlight the progressive stance of its environmental law, a position that must be consistently reaffirmed to avoid the threat of regression driven by powerful economic interests incompatible with the objectives of the Paris Agreement and the 2030 Agenda for Sustainable Development. Despite its broad protections, grounded in constitutional and special laws, Brazil's environmental law—still anthropocentric—evolves amidst both advances and setbacks. It contends with denialist ideologies, utilitarian and individualistic philosophies, and actors seeking to enshrine unsustainable, short-term perspectives. Beyond laws and courts, Brazil must, rooted in the principle of environmental education, develop ethical and philosophical foundations to counter the powerful economic, social, and political interests often cloaked in the seductive promises of immediate financial gain. These interests, however, are evident adversaries of the fundamental values of the Environmental Rule of Law.

In this study, the authors address with clarity and precision the critical themes of Brazilian environmental law: its principles; environmental constitutionalism; animal law; the allocation of competencies among federal entities; the National Environmental Policy; the bodies and entities of the National Environmental System; administrative instruments for environmental regulation, particularly police power; administrative liability; licensing; environmental impact assessment; zoning; civil and criminal liability; the role of the judiciary in implementing environmental norms; specially protected areas; climate law; and the National Water Resources Policy.

The authors, as Federal Judges, provide a detailed analysis of Brazil's legal system and the jurisprudence of the country's Superior Courts. Their work demonstrates a clear commitment to the maximal effectiveness and realization of Article 225 of the Constitution and the environmental protection laws necessary to rigorously combat pollution and global warming.

Superior Court of Justice,
Brasilia, Brazil
4 November 2024

Reynaldo Soares da Fonseca

Acknowledgments

We would like to extend our gratitude to Thiago Calmon for his meticulous review of the translation of this work into English. His expertise and attention to detail have significantly enhanced the quality and accuracy of the translation.

Thiago Gomes Calmon is a distinguished professor of Legal English with extensive experience teaching at esteemed institutions across Brazil. Throughout his career, he has collaborated with prestigious law firms, judicial associations, and universities, offering specialized instruction in international contracts and legal writing. His academic contributions include professorships at the Superior Tribunal de Justiça (STJ), Escola Paulista de Direito (EPD), and the Associação dos Juízes Federais do Brasil (AJUFE), among other renowned institutions.

In addition to his teaching endeavors, Professor Calmon is a published author, having written books on legal English writing and the language of international contracts.

Competing Interests The authors have no competing interests to declare that are relevant to the content of this manuscript.

Contents

1 Introduction 1

2 Environmental Law in Brazil: Historical Path and Current Outlook 7
- 2.1 Introduction 7
- 2.2 Historical Evolution of Environmental Law in Brazil 9
- References 12

3 Principles of Environmental Law 15
- 3.1 Introduction 15
- 3.2 Principle of Sustainability or Sustainable Development 15
- 3.3 Principle of the Socio-Environmental Function of Property 19
- 3.4 Polluter-Pays Principle 20
- 3.5 Prevention and Precautionary Principle 22
- 3.6 Principle of Publicity, Transparency, or Access to Information in Environmental Matters 26
- 3.7 Principle of Public Participation 28
- 3.8 Principle of Non-Regression 32
- 3.9 Principle of *In Dubio Pro Natura* 34
- References 35

4 Environmental Constitutionalism in Brazil 39
- 4.1 Introduction 39
- 4.2 Environmental Law in the Brazilian Constitution 40
- 4.3 Environmental Protection and Indigenous Peoples 42
- 4.4 Legal Regime of Environmental Assets in Brazilian Law 44
- References 46

5 Animal Protection and Animal Rights 49
- 5.1 Introduction 49
- 5.2 Animal Protection in Brazilian Law 50
- References 53

6 Environmental Federalism 55
6.1 Introduction 55
6.2 Material or Administrative Competence 56
6.2.1 Exclusive Administrative Competence of the Federal Government 57
6.2.2 Remaining or Residual State Authority 57
6.2.3 Administrative Competence of Municipalities 57
6.2.4 Federative Cooperation and the Supplementary Law n. 140/2011 58
6.3 Legislative Competence 59
6.3.1 Exclusive Legislative Competence of the National Congress (Federal Government) 59
6.3.2 Remaining or Residual Legislative Competence of the States 60
6.3.3 Concurrent Legislative Competence 60
6.3.4 Supplementary Legislative Competence of Municipalities 64
References 66

7 National Environmental Policy and National Environmental System 67
7.1 Introduction 67
7.2 National Environmental System (SISNAMA) 68
7.2.1 National Council for the Environment (CONAMA) 69
7.2.2 Authorities Responsible for Implementing Environmental Legislation 72
7.3 The Role of other Public Administrative Bodies and Regulatory Agencies 73
References 74

8 Administrative Tools for Environmental Regulation 75
8.1 Introduction 75
8.2 Administrative Enforcement of Environmental Law: Administrative Violations and Penalties 76
8.2.1 Legal Regime of Administrative Liability 76
8.2.2 The Principle of Legality 78
8.2.3 Liable Party 79
8.2.4 Enforcement Authorities 80
8.2.5 Duty of Surveillance by the Public Authorities 81
8.2.6 Administrative Penalties Against Environmental Violations 82
8.2.7 Due Administrative Process, Time Limitations, and Judicial Review 83
8.3 Environmental Licensing and Environmental Impact Assessment 85

8.3.1 Activities Subject to Environmental Licensing 88
8.3.2 Federative Competence for Environmental Licensing 89
8.3.3 Types of Environmental Licenses 94
8.3.4 Validity Periods of Licenses and Deadlines for Decisions on License Applications 97
8.3.5 License Amendment and Withdrawal 98
8.3.6 Environmental Studies and Environmental Impact Assessment (EIA) 99
8.4 Land Use Regulation and Environmental Zoning 104
8.5 Economic Tools for Environmental Regulation 107
8.5.1 Environmental Taxation 107
8.5.2 Payment for Environmental Services 109
References 111

9 Environmental Civil Liability 113
9.1 Introduction 113
9.2 Legal Framework for Environmental Civil Liability 114
9.3 Civil Liability for Environmental Harm: Requirements 115
9.3.1 Health and Environmental Risk Activity 116
9.3.2 Cause and Effect Correlation 117
9.3.3 Environmental Risk and Harm 118
9.3.4 Damages for Collective Nonpecuniary Loss 120
9.3.5 Destination of the Indemnification Amount 123
9.3.6 State Liability for Environmental Damage 123
References 124

10 Environmental Criminal Liability 127
References 131

11 Environmental Law in Court 133
11.1 Introduction 133
11.2 Environmental Courts 134
11.3 Class Actions and Public Interest Litigation in Brazil 136
11.4 Legal Lawsuits for Environmental Protection 137
11.4.1 Public Civil Suit 138
11.4.2 People's Legal Suit 142
11.4.3 Constitutional Suits 142
11.5 Judicial Deference and the Scope of Judicial Review 143
References 146

12 Specially Protected Areas 149
12.1 Introduction 149
12.2 Nature Conservation Units' System 151
12.3 Forest Protection Code 153
12.3.1 Main Features of the 2012 Forest Protection Code 154

12.3.2 Permanent Preservation Areas 156
12.3.3 Rural Property Preservation Area 160
12.3.4 Regulating the Use of Fire in Farming Activities 162
12.3.5 Regularization of Non-Compliant Activities 163
References 164

13 Climate Change Law and Climate Change Litigation 167
13.1 Introduction 167
13.2 The Awakening to a New Crisis: The Green House Effect 168
13.3 The International Response to the Climate Crisis 170
13.4 Brazilian Law on Climate Change 176
13.5 Brazilian Greenhouse Gas Emissions Trading System 179
13.6 Climate Litigation in Brazil 182
References 188

14 National Water Resources Policy 191
14.1 Introduction 191
14.2 Constitutional and Legal Protection of Water Resources 192
14.3 Administrative and Legislative Competence 193
14.4 National Policy on Water Resources 195
14.5 Allocation of Water Resources 196
14.6 Charging for the Use of Water Resources 197
14.7 Water Resource Use Regulation 197
Reference 198

15 Conclusion 199

Glossary 203

About the Author

Rafael Martins Costa Moreira is a federal judge in the 2nd Court of Canoas/RS and holds a degree in Legal and Social Sciences from the Pontifical Catholic University of Rio Grande do Sul (PUCRS) (2002). Master's (2016) and PhD (2020) in law from PUCRS, with a period of visiting research at Ruprecht Karls Universität Heidelberg, Germany. Professor of Environmental and Administrative Law. President of AJUFERGS (Association of Federal Judges of Rio Grande do Sul), 2020–2022 and 2024–2026 biennia. Has experience in law, with an emphasis on administrative and environmental law.

Gabriel Wedy is a federal judge and a professor of the Graduate Program and at the School of Law at the Universidade do Vale do Rio dos Sinos (UNISINOS) and at the School of the Federal Judiciary (ESMAFE). He holds Post-Doctoral, Doctoral, and Master's degrees in law. Professor Wedy is a member of the World Commission on Environmental Law (IUCN-WCEL) and the European Law Institute (ELI). He is a vice-president of Law for a Green Planet Institute (IDPV), and former president of the Association of Federal Judges of Brazil (AJUFE). He is a former visiting scholar at Columbia Law School (Sabin Center for Climate Change Law) and at Universität Heidelberg (Instituts für deutsches und europäisches Verwaltungsrecht).

Abbreviations

APP	Área de Preservação Permanente (Permanente Preservation Area)
COP	Conference of the Parties
EIA	Estudo de Impacto Ambiental (Environmental Impact Assessment)
GDP	Gross Domestic Product
GHG	Greenhouse Gas
GMO	Genetically Modified Organism (Organismo Geneticamente Modificado - OGM)
HDI	Human Development Index
PL	Projeto de Lei (bill)
RIMA	Relatório de Impacto Ambiental (Environmental Impact Report)
SINARE	Sistema Nacional de Redução de Emissões (National System for the Reduction of GHG Emissions)
SISNAMA	Sistema Nacional do Meio Ambiente (National Environmental System)
UNESCO	United Nations Educational, Scientific and Cultural Organization

Chapter 1
Introduction

This book functions as a guide to Brazilian environmental and climate change law for an international audience. It is designed for students, lawyers, jurists, scholars, and anyone interested in exploring the key features and current landscape of Brazilian environmental law.

Given that Brazil's environmental landscape has attracted international attention and considering the scarcity of English-language literature on the subject (along with the challenge many foreign scholars face in reading Portuguese), this book was crafted to provide accessible insights for those interested in how laws and the justice system address environmental disputes in this country.

There are numerous compelling reasons to delve into Brazilian environmental and climate change law. Beyond its sheer expanse and populous nature, Brazil cradles within its borders the Amazon rainforest—a vast and ancient green world, the largest tropical forest on Earth, a sanctuary of extraordinary biodiversity, and the wellspring of myriad ecosystem services. The Amazon stands as a formidable guardian of global ecology, sheltering 40% of the planet's remaining rainforests, harboring a quarter of its terrestrial biodiversity, and nurturing a wealth of fish species unmatched by any other river system. Yet, the Amazon, alongside Brazil's other vital forests, holds another role of profound importance: that of a carbon sink, a mantle it wears with dignity, though one now threatened by the unrelenting forces of deforestation and climate change.

Environmental concerns have profoundly shaped Brazil's international trade relationships. For example, the European Union-Mercosur Trade Agreement hinges on Brazil's environmental commitments, with EU lawmakers pressing for stringent environmental safeguards. Recognizing Brazil's global ecological significance, the World Bank's board of directors recently approved a $500 million project to enhance sustainability-linked financing, bolster the private sector's capacity in carbon credit markets, and support deforestation reduction efforts. Additionally, countries such as Germany, Norway, and Japan invest substantial funds in Brazil's environmental preservation programs.

R. Martins Costa Moreira, G. Wedy, *Brazilian Environmental and Climate Change Law*, https://doi.org/10.1007/978-3-031-87998-2_1

Beyond "forest treasure," Brazil has taken a relevant position in the control of greenhouse gas emissions and in the struggle against global warming. Brazil's greenhouse gas emissions rose more than 12% in 2021, largely due to surging deforestation in the Amazon rainforest. Data shows that in 2021 the country emitted 2.42 billion tons of carbon dioxide equivalent (CO2e), an increase from the 2.16 billion tons of 2020. Yet, as unveiled at COP 28 in December 2023, data revealed an 8% decline in emissions from 2021, a decrease chiefly attributed to reduced reliance on fossil fuels for electricity generation. In 2022, Brazil emitted 2.3 billion tons of greenhouse gases. Despite this significant reduction, emissions for that year still soared to the third highest level since 2005, trailing only behind the peaks of 2019 and 2021.

Despite its relatively clean electricity grid, which leans heavily on hydropower and renewable sources, Brazil stands as the world's sixth-largest emitter of greenhouse gases. This paradox arises largely from its deforestation, agricultural practices, and various forms of land use.

Brazil is also connected to other environmentally significant sectors: its invaluable water resources, numerous indigenous communities, the richest biodiversity on the planet, and its status as one of the largest waste producers globally. Brazil also boasts an extensive coastal zone, ranks as one of the world's foremost food producers, leads in hydropower usage, and holds immense potential for renewable energy development, among other attributes. It is clear, therefore, that how Brazil manages its environment, ecosystems, and emissions will have profound global implications. In other words, it is appropriate to describe Brazil as an environmental, forest, climate, water, and energy "superpower."

Furthermore, Brazil has much to contribute to the international discourse on environmental law. While enforcement remains a challenge, Brazil possesses an advanced and protective legislative framework. The duty to protect the environment is prominently enshrined in the 1988 Brazilian Constitution (Art. 225). Although the country lacks a single statute or "environmental law code," it has established sophisticated, detailed regulations across various sectors of environmental law. Additionally, Brazil has well-structured public institutions—such as the Public Prosecutor's Office, the Public Defender's Office, the Attorney General's Office, and the Courts of Auditors, along with environmental bodies and agencies—and an independent Judiciary capable of enforcing environmental legislation and holding polluters accountable for environmental damage.

To provide academics, lawyers, and international readers with an understanding of the main contours of Brazilian environmental law, this book will cover the fundamental and most pertinent topics.

We begin our exploration of Brazilian environmental law with an overview of its historical evolution and current landscape (Chap. 2), highlighting the principles of environmental law (Chap. 3). Notably, in Brazil, legal principles hold significant weight in judicial reasoning, as judges and courts can impose implicit duties on public authorities and private individuals based on constitutional provisions. These principles, particularly those related to environmental protection, serve as a foundation for establishing such duties.

Special attention is given to Brazilian environmental constitutionalism (Chap. 4), emphasizing the pioneering 1988 Constitution, which established explicit environmental protection duties for both the public sector and society. The Constitution reinforces the status of environmental resources as common goods, belonging to society as a whole, alongside the recognition of animal rights within our legal system (Chap. 5).

A substantial chapter is dedicated to *environmental federalism* (Chap. 6), examining the distribution of administrative and legislative powers between the federal government and subnational entities. In large, federal countries like Brazil, the allocation of authority among federative entities and the ensuing conflicts are of particular significance.

The National Environmental Policy and the National Environmental System, established by Law n. 6.938/81, are examined in depth (Chap. 7). Law n. 6.938/81 is considered the foundation of Brazilian environmental law, marking the first time environmental issues were addressed systematically. This statute outlines key concepts and tools of environmental policy, along with the bodies and agencies tasked with the formulation and implementation of environmental law.

Chapter 8 explores the administrative and regulatory tools for environmental protection. It begins by highlighting the traditional command-and-control approach, which involves the administrative enforcement of environmental laws, including violations and penalties. This is followed by an analysis of the environmental license and Environmental Impact Assessment. The environmental license is a pivotal mechanism in Brazilian environmental law, often regarded as the "heart of environmental control," as all activities that cause or may cause environmental impacts must undergo prior licensing from the appropriate environmental protection agency. Within this licensing process, activities with potentially significant environmental impacts also require an Environmental Impact Assessment before a license can be issued. Additionally, land use regulation and environmental zoning are briefly discussed. Lastly, economic tools for environmental regulation are examined, with a focus on environmental taxation and payment for environmental services.

Chapter 9 addresses environmental civil liability. It emphasizes that failing to hold industries accountable for the costs they impose on the environment would effectively amount to a "gift" for polluters and exploiters of environmental resources, allowing them to benefit from common goods without fair compensation to society. Compensation for environmental harm is typically pursued through class actions in the Judiciary, with the "public civil suit" being the primary legal instrument used in Brazil for such claims.

We also explore environmental criminal liability (Chap. 10). The 1988 Constitution (Art. 225, § 3) addresses responsibility for environmental crimes. Law n. 9.605/98 specifically defines various environmentally harmful acts as crimes. Remarkably, under this legislation, legal entities may be prosecuted as defendants in criminal proceedings, but only for environmental offenses.

Chapter 11 examines environmental law in the courts. Although Brazil follows a "civil law system," jurisprudence has a distinctive role in shaping Brazilian law. Brazilian judges enjoy notable independence compared to their counterparts abroad, and the Judiciary holds prerogatives designed to shield it from political and economic pressures. This grants the courts a vital role in resolving environmental disputes and safeguarding ecosystems. Besides, the legal system has developed various procedural mechanisms to ensure that environmental harms and threats are brought before the Judiciary.

Chapter 12 delves into specially protected areas, given their environmental importance. Protected areas, particularly forests, are of special interest, as Brazil's land use, land-use change, and forestry sectors significantly drive climate change. Two primary pieces of legislation address this issue in Brazil: Law n. 9.985/00, which governs Nature Conservation Units, and Law n. 12.650/12, known as the "Forest Protection Code." The former establishes rules for creating and regulating protected or conservation areas, with specific limitations depending on the purpose of each unit. The latter addresses one of the most contentious issues in Brazilian law: the regulation of native vegetation protection and land use. The Forest Protection Code emerged from extensive debate in the National Congress, and its implementation has sparked numerous conflicts requiring judicial resolution.

Here we turn to the topic of climate change law and climate litigation (Chap. 13). As the United Nations has stated, climate change is the defining issue of our time, and we are at a defining moment. Brazil, as the world's sixth largest emitter of greenhouse gases, has much to help mitigate global warming. However, climate change law and climate litigation in Brazil remain in an early stage and have yet to achieve significant effectiveness.

Although Brazil has recently seen a slight decline in emissions, the overall trend remains upward. The country still lacks clear targets for both public and private sectors and has not yet established a comprehensive carbon market or systematic carbon tax framework. Nevertheless, there have been some advancements. Brazil has a foundational statute on Climate Policy (Law n. 12.187/2009), and new measures focusing on renewable energy, carbon markets, and carbon taxation are currently under development in the National Congress and the Executive Branch. Additionally, Brazil benefits from an independent Judiciary and robust institutions, such as the Public Prosecutor's Office and the Public Defender's Office, which are well-positioned to pursue climate litigation, aiming to hold public authorities accountable for protective measures or to seek compensation for climate-related damages.

Finally, the reader is presented with a study on water resource regulation in Brazil (Chap. 14). Brazil holds a prominent position globally in terms of freshwater resources, containing 12% of the planet's reserves and accounting for 53% of South America's water resources. Transboundary river basins cover 60% of Brazilian territory. However, the country has seen a 15.7% reduction in surface water since the early 1990s, from nearly 20 million hectares to 16.6 million hectares in 2020. Brazil is also home to three of the world's largest aquifers: the Amazon Basin, the Guarani Aquifer System, and the Maranhão Basin.

This book, therefore, takes on the challenge of explaining a complex subject filled with technical and specialized terms in a clear, didactic, and accessible language, avoiding excessive "legalese." The goal is to present the main aspects and current landscape of Brazilian environmental and climate change law, drawing on existing legislation, jurisprudence from the Supreme Court and the Superior Court of Justice, and relevant environmental law literature.

Chapter 2
Environmental Law in Brazil: Historical Path and Current Outlook

2.1 Introduction

The commitment to ensuring sustainable development, fostering intergenerational solidarity, and promoting environmental justice has firmly entered the agendas of countries and international organizations, becoming a central topic in discussions about the future of humanity. Conflicts over natural resource exploitation and air, water, and soil contamination have long existed, but today they are increasingly central to sustainable development, human well-being, and the very survival of humanity. We live in a global community marked by invisible risks that affect our health and ecosystems—risks that consistently push beyond the planet's resilience limits.

In regard to the climate emergency, scientific findings increasingly point to the anthropogenic origins of global warming, triggered since the Industrial Revolution. This human-driven warming has led to harmful effects, including shrinking glaciers, rising sea levels, warming ocean temperatures, heightened climate-related risks, and numerous associated impacts.[1]

The end of the twentieth century can be considered as the era of humanity's awakening to the dangers to its very survival[2] due to human activity itself, the emergence of the risk society,[3] and the current period that some geologists have coined the *Anthropocene*, that is, the physical transformation of the earth by human action.[4] Despite its rejection as a formal unit of the Geologic Time Scale,

[1] On the recent evidence of the seriousness of global warming and climate change caused by human action and the measures needed to contain its impacts, see the latest IPCC report (AR6): IPCC (2021).

[2] According to Juarez Freitas, everything indicates that in the next million years, the planet will not become extinct. It is humanity that is in real danger (Freitas 2019, p. 25).

[3] Beck (1998).

[4] Crutzen and Stoermer (2000).

R. Martins Costa Moreira, G. Wedy, *Brazilian Environmental and Climate Change Law*, https://doi.org/10.1007/978-3-031-87998-2_2

Anthropocene will nevertheless continue to be used not only by Earth and environmental scientists, but also by social scientists, politicians and economists, as well as by the public at large. It will remain an invaluable descriptor of human impact on the Earth system.[5]

This reality raises the question of overcoming the "planetary boundaries,"[6] beyond which human activities can propel the planet towards unknown and dangerous situations of climatic imbalance, biodiversity loss, and change in the chemical composition of the air, land, and oceans.[7]

The Brazilian legal system has also been influenced by the development of environmental law globally: the Stockholm Declaration of 1972, the Brundtland Report of 1987, and the Rio Declaration of 1992, to the most recent international acts such as the Johannesburg Declaration of 2002 and Rio Declaration of 2012 (*Rio follow-up process*), the Paris Agreement of 2015 or Conference of Parties (COP) 21, and the 2030 Agenda for Sustainable Development.

In Brazil, the legal protection of the environment has been through some gradual development. Nowadays, we have not only the provision of a fundamental right to an ecologically balanced environment under our Constitution, but also a broad regulatory framework and statutory provisions of environmental law. However, despite the large amount of legislation on the subject, and even its good quality level, Brazil has suffered, like much of the world, from a crisis of effectiveness in environmental law enforcement. That is to say, environmental destruction is accelerating even after the enactment of environmental laws.[8]

This "crisis of effectiveness"[9] is quite pronounced in Brazil. This is evidenced by the low payment rate of environmental fines, flaws in the administrative process,[10] significant reduction in enforcement, and an increase in illegal deforestation and wildfires, albeit with little human development.[11] It is worth mentioning that deforestation in the "Legal Amazon" increased by 21.97% in 2021,[12] and wildfires in the

---

[5] IUGS (2024).

[6] About the nine resilience limits of the planet, within which humanity can develop in a sustainable way, without abrupt and irreversible environmental changes, see: Stockholm Resilience Center (2023).

[7] Along those lines: Sachs (2015).

[8] Cullinam (2013), p. 95.

[9] Vaz (2009).

[10] The shortcomings in an environmental administrative procedure conducted by the federal agency IBAMA (*Instituto Brasileiro do Meio Ambiente e dos Recursos Naturais Renováveis* or *Brazilian Institute for the Environment and Renewable Natural Resources*. See more in Sect. 7.2.2), mainly related to the slowness of the process, including the risk of termination by the mere lapse of time (*statute of limitations*), and the very low payment rate of the fines imposed, are described in a report prepared by governmental authorities (Brazil 2017, 2019). For a critical analysis of these data, see: Libório (2019).

[11] As stated by Brazilian Supreme Federal Court justice ("Ministro") Luís Roberto Barroso, after the public hearing held to instruct the judgment of ADPF 708, which dealt with the operation of the National Fund on Climate Change (Climate Fund) (Brazil 2020b. STF).

[12] Agência Brasil (2021).

Pantanal increased by almost 200% between 2019 and 2020.[13] In 2022, the Amazon suffered a fifth consecutive record year in deforestation, representing an increase of almost 150% over the previous quadrennium.[14] In January and February 2023, it was identified 489 km^2 of area with signs of deforestation in the Legal Amazon. This is the second largest area deforested in these two months since 2016, the beginning of the historical series.[15]

According to a well-known statement by the Superior Court of Justice Judge Herman Benjamin about the current state of Brazilian environmental law, in Brazil, unlike other countries, a judge does not need to create laws for the protection of the environment. Environmental obligations flow from the legislative texts. Fortunately, he says

> our Judiciary is not haunted by an ocean of gaps or a festival of legislative half-words. If a gap exists, it is not for lack of legislation, or even a defect in the statute; it is for the absence or deficiency of administrative and judicial implementation of the unequivocal environmental duties established by the legislation.[16]

It is important, then, to sketch an overview of the historical evolution of Brazilian environmental law, and then report on the current situation.

2.2 Historical Evolution of Environmental Law in Brazil

The development of Brazilian environmental law is the result of a slow legislative evolution, with emphasis on the provision of environmental rules in the 1988 Constitution, without losing sight of the growing contribution of the courts' jurisprudence. For didactic purposes, according to Benjamin, we may identify three stages in the history of Brazilian environmental law: (a) the stage of *unregulated exploitation*; (b) the *fragmentary stage*; and (c) the *holistic stage*.[17]

The stage of *unregulated exploitation* or *environmental laissez-faire*, which took place from 1500 (the date of the "discovery" of Brazil) until the beginning of the second half of the twentieth century, was characterized by the absence of genuine environmental concern, with isolated rules aimed mainly to ensure the maintenance of some precious natural resources in an accelerated process of depletion and to safeguard human health. Environmental conflicts were decided in an unsystematic way, through private law and neighborhood lawsuits.

In the *fragmentary stage* the legislator—now concerned with broad categories of natural resources, but not yet with the environment as such—imposed legal restrictions on certain exploitation activities. These include the Forest Protection Act of

[13] Brazil (2020a). Câmara dos Deputados.

[14] Imazon (2023).

[15] Brazil (2023).

[16] Superior Court of Justice, REsp 650.728/SC, Rapporteur Min. Herman Benjamin, j. 23 Oct. 2007.

[17] These stages are reproduced from the article: Benjamin (1999).

1965; the Hunting, Fishing, and Mining Acts, all of 1967; the Liability for Nuclear Damage Act, of 1977; the Industrial Zoning in Critical Pollution Areas Act, of 1980; and the Agrochemicals Act, of 1989. Of course, these cycles are not completely separate, but they overlap for a certain timeframe until the consolidation of the new phase.

Finally, the *holistic stage* began with the enactment of the National Environmental Policy Act (Law n. 6.938/81), in which the environment was protected as a whole, that is, as an integrated ecological system with its own, autonomous legal value. It was with Law n. 6.938/81, hence, that environmental protection has truly begun in Brazil. This Act sets forth the principles, objectives, and legal instruments of Brazilian's National Environmental Policy, laying down the environmental impact assessment into the Brazilian legal system, establishing a regime of strict liability (regardless of fault) for environmental damage, and vested the Public Prosecutor's Office with standing to take legal action to this purpose for the first time.[18]

This chronological arrangement can be organized in another way. In the view of Fensterseifer and Sarlet, one can identify a *first fragmentary-instrumental phase*, prior to Law n. 6.938/81; a *second systematic-evaluative phase*, beginning with Law n. 6.938/81, which gave birth to environmental law in Brazil; and the *third phase* of *constitutionalizing* the environmental protection, under the 1988 Constitution.[19]

The foundation of Brazilian environmental law, as experienced in other countries, was influenced by the international environmental acts and environmental movement, still in the 1970s.[20] But the onset of an *ecological shift* in Brazilian legislation took place in the early 1980s, with the enactment of the National Environmental Policy Act (Law n. 6.938/81) and was further strengthened by the constitutional enshrinement of ecological rights and values in the Brazilian Constitution of 1988.[21]

The National Environmental Policy Act (Law n. 6.938/81) has indeed marked a crucial moment in the evolution of Brazilian environmental law, followed by the Public Civil Suit Act of 1985 (Law n. 7.347/85)[22] and CONAMA[23] regulation on licensing and environmental impact assessment (which will be further analyzed later in this book). However, the great milestone that Brazilian law has reached to embrace environmental protection as a duty of the State and society was the 1988 Constitution. It is considered by many as an *Ecological Constitution* or *Green*

[18] Benjamin (1999).

[19] Fensterseifer and Sarlet (2021), p. 145.

[20] On the evolution of the environmental movement and ecological awareness in Brazil, one can consult: Fensterseifer and Sarlet (2019), pp. 85–94; de Faria Tavolaro (2001); Viola (1987).

[21] Fensterseifer and Sarlet (2019), p. 135.

[22] The Public Civil Suit Act of 1985 (Lei n. 7.347/85), along with the Consumer Protection Code (Lei n. 8.078/90), are the main legislative acts that regulate class actions under Brazilian law, which can also be filed for environmental protection, as will be better studied bellow.

[23] CONAMA is short for the "National Environmental Council." It is a federal representative body, with the function of issuing environmental regulations that are hierarchically inferior to legislation. We shall deal with CONAMA in Sect. 7.2.1.

Constitution.[24] As we shall see, the constitutional provisions for the environment, especially by requiring that environmental issues should be considered prior to social and economic development and upon government decision-making, as well as the provisions within the same scope as fundamental rights, have come to influence the enactment of new pieces of legislation, the performance and organization of the Executive branch, and the interpretation of the courts. The entire environmental legislative framework that has emerged after 1988 is guided and limited by constitutional rules.

It is important to note that the Brazilian Constitution was ratified after a long period of military dictatorship (which had begun in 1964), and that the drafters of the 1988 Constitution intended to provide institutional safeguards for democracy, such as the separation of powers, the strengthening of the Judiciary's and the public prosecutor's independence, and the promotion of fundamental rights; besides, apart from these provisions, it has also stipulated rules for social and environmental rights.

After the Constitution, then, several legislative acts were created to regulate specific issues of environmental law: Water Resources Act of 1997 (Law n. 9.433/97); Environmental Crimes Act of 1998 (Law n. 9.605/98, which also contains provisions on the administrative sanctioning process); Nature Conservation Units Act of 2000 (Law n. 9.985/00); Biosafety Act of 2005 (Law n. 11.105/05); Public Forest Management Act of 2006 (Law n. 11.284/06); Atlantic Forest Act of 2006 (Law n. 11.428/06); Climate Change Act of 2009 (Law n. 12.187/09); Solid Waste Act of 2010 (Law n. 12.305/10); the cooperation Act between federal, states, Federal District, and local governments in administrative actions on environmental matters (Supplementary Law n. 140/11); the Forest Protection Act of 2012 (Law n. 12.651/12); Biodiversity Protection Act of 2015 (Law n. 13.123/15); the National Policy for Payment for Environmental Services Act (Law n. 14.119/21), the new Pesticide Control Act (Law n. 14.785/23); the Climate Change Adaptation Plans Act (Law n. 14.904/24), to mention the most relevant legislation.

There is still no specific federal legislation about environmental licensing. As shall be mentioned below, environmental licensing is basically regulated by CONAMA's regulations, and State and Municipal rules. However, there is a bill waiting to be voted in the National Congress on this matter.[25]

In Brazil, as across the globe, there is a rising tide of awareness and concerted efforts toward environmental protection. However, there is a fierce ideological debate between those who defend economic growth as a priority, especially because of economic crises, worsened after the COVID-19 pandemic and the rise in prices in general (mainly due to Russia's war on Ukraine), and those who defend the necessity and urgency of preserving ecosystems and ecological balance. No case can be more representative of this conflict than the recent debate about oil exploration at the

[24] Fensterseifer and Sarlet (2019), p. 55.

[25] This bill was approved by the House of Representatives (*Câmara dos Deputados*) in 2021, and it is currently being analyzed by the Senate, under the n. 2.159-21 (Bill ("*Projeto de Lei*" or PL) n. 2.159/21).

mouth of the Amazon River. This issue places sectors of the government at odds with one another: the state-owned oil company Petrobras and the Executive branch advocate for exploration, while the environmental agency IBAMA and the Ministry of the Environment oppose the venture. These clashing perspectives are as enduringly familiar as ever: economic development for the Amazon region and Brazil *versus* ecological and water protection.[26]

The pressure to alleviate environmental regulation is not recent in Brazilian political history and has occurred during governments of different shades of opinion and ideologies.[27] Some governments embraced a perspective more favorable to economic growth than others. In Brazil, however, environmental protection faces significant obstacles within the political sphere, making it frequently necessary to turn to the Judiciary, the Public Prosecutor's Office, and other independent institutions to strengthen ecological protection.

That is to say, it will be up to organized civil society and independent authorities, such as the Public Prosecution, the Public Defender's Office, and the Audit Courts, supported by an independent and impartial Judiciary, to confront abusive conduct, by action or omission, that goes against the legal, constitutional, and international framework for environmental protection. All of this, naturally, does not preclude the need for an essential modernization of environmental legislation.

References

Agência Brasil (2021) INPE: desmatamento na Amazônia Legal tem aumento de 21,97% em 2021. https://agenciabrasil.ebc.com.br/geral/noticia/2021-11/desmatamento-na-amazonia-legal-tem-aumento-de-2197-em-2021. Accessed 18 Nov 2024

Beck U (1998) Weltrisikogesellschaft. Suhrkamp, Frankfurt am Main

Benjamin AHV (1999) Introdução ao direito ambiental brasileiro. Revista de Direito Ambiental 14: 48–82

Brazil (2017) Ministério do Meio Ambiente. IBAMA. *Relatório de Gestão – 2017*. https://www.ibama.gov.br/auditorias/relatorios. Accessed 18 Nov 2024

Brazil (2019) Controladoria-Geral da União (CGU). CGU avalia desempenho e gestão do processo sancionador ambiental do IBAMA. https://www.cgu.gov.br/noticias/2019/04/cgu-avalia-desempenho-e-gestao-do-processo-sancionador-ambiental-do-ibama. Accessed 18 Nov 2009

Brazil (2020a) Câmara dos Deputados. Inpe confirma aumento de quase 200% em queimadas no Pantanal entre 2019 e 2020. https://www.camara.leg.br/noticias/696913-inpe-confirma-aumento-de-quase-200-em-queimadas-no-pantanal-entre-2019-e-2020/. Accessed 18 Nov 2024

Brazil (2020b) STF. Barroso defende agenda efetiva de proteção ambiental no encerramento da audiência pública. http://portal.stf.jus.br/noticias/verNoticiaDetalhe.asp?idConteudo=452174&ori=1. Accessed 18 Nov 2024

[26]To better understand this clash, read on: Financial Times (2023).

[27]For an analysis of proposals for environmental deregulation in Brazil involving several political groups, it is worth consulting: da Silva Barcelos (2020).

Brazil (2023) Ministério do Meio Ambiente e Mudança do Clima. Análise dos alertas de desmatamento na Amazônia Legal: jan/fev 2023. https://www.gov.br/mma/pt-br/analise-dos-alertas-de-desmatamento-na-amazonia-legal-jan-fev-2023. Accessed 18 Nov 2024

Crutzen PJ, Stoermer EF (2000) The "Anthropocene". Global Change Newsletter 41:17–18. http://www.igbp.net/download/18.316f18321323470177580001401/1376383088452/NL41.pdf. Accessed 18 Nov 2024

Cullinam C (2013) The rule of Nature's law. In: Voigt C (ed) Rule of law for nature: new dimensions and ideas in environmental law. Cambridge University Press, New York, pp 94–108

da Silva Barcelos EÁ (2020) Desregulação ambiental e disputas políticas: Uma breve retrospectiva do desmonte do licenciamento ambiental no Brasil. AMBIENTES: Revista de Geografia e Ecologia Política 2-2:278

de Faria Tavolaro SB (2001) Movimento ambientalista e modernidade: sociabilidade, risco e moral. Annablume/Fapesp, São Paulo

Fensterseifer T, Sarlet IW (2019) Direito Constitucional Ecológico, 6th edn. Revista dos Tribunais, São Paulo

Fensterseifer T, Sarlet IW (2021) Curso de Direito Ambiental, 2nd edn. Forense, Rio de Janeiro

Financial Times (2023) Brazil divided over oil drilling proposal off Amazon delta, jun. 8 2023. https://www.ft.com/content/800693a0-fc52-408e-83bd-0b8e63f38b3a. Accessed 18 Nov 2024

Freitas J (2019) Sustentabilidade: Direito ao Futuro, 4th edn. Forum, Belo Horizonte

Imazon (2023) Amazônia perdeu quase 3 mil campos de futebol por dia de floresta em 2022, maior desmatamento em 15 anos. https://imazon.org.br/imprensa/amazonia-perdeu-quase-3-mil-campos-de-futebol-por-dia-de-floresta-em-2022-maior-desmatamento-em-15-anos/. Accessed 18 Nov 2024

IPCC (2021) Climate Change 2021. The Physical Science Basis. Summary for Policymakers. https://www.ipcc.ch/report/ar6/wg1/downloads/report/IPCC_AR6_WGI_SPM_final.pdf. Accessed 18 Nov 2024

IUGS. International Union of Geological Sciences (2024) The Anthropocene: IUGS-ICS Statement, 20 Mar. 2024. https://www.iugs.org/_files/ugd/f1fc07_23c6f9e723bc47b9b5fdcd300f806f25.pdf?index=true. Accessed 18 Nov 2024

Libório B (2019) Por que o Ibama arrecada só 5% das multas ambientais que aplica. Aos fatos. https://aosfatos.org/noticias/por-que-o-ibama-arrecada-so-5-das-multas-ambientais-que-aplica/. Accessed 18 Nov 2024

Sachs J (2015) The age of sustainable development. Columbia University Press, New York

Stockholm Resilience Center (2023) Sustainability Science for Biosphere Stewardship. Planetary boundaries research. http://www.stockholmresilience.org/research/planetary-boundaries.html. Accessed 18 Nov 2024

Vaz PAB (2009) Direito administrativo ambiental: aspectos de uma crise de efetividade. Interesse Público- IP 11(56):115–157

Viola EJ (1987) O movimento ecológico no Brasil (1974-1986): do ambientalismo à ecopolítica. Revista Brasileira de ciências sociais 1(3):5–26

Chapter 3
Principles of Environmental Law

3.1 Introduction

Without delving into the theoretical nuances and various controversies surrounding the issue, we can affirm that the principles of environmental law are standards aimed at protecting the fundamental right to an ecologically balanced environment and providing interpretive guidance. In Brazil, legal principles hold particular importance in judicial reasoning, as judges and courts can impose implicit duties on public authorities and private individuals based on constitutional provisions. These principles establish duties, especially concerning environmental protection, and assist the courts in interpreting legislative acts.

Environmental law literature categorizes these principles in various ways, though there is consensus on the existence of some of them. This book addresses only the most widely recognized and specific principles of environmental law in Brazil: sustainable development (or sustainability); the social and environmental function of property; polluter-pays principle; prevention and precautionary principle; publicity; public participation; non-environmental regression; and *in dubio pro natura* (in doubt, in favor of nature).

3.2 Principle of Sustainability or Sustainable Development

The principle of sustainability or sustainable development emerged mainly in the last two decades of the twentieth century, especially due to international agreements, such as the Brundtland Report of 1987, the Rio Declaration of 1992, and the Rio-follow-up process (that is, the Rio Conference of 2012 or Rio+20).[1]

---

[1] Gärditz and Kahl (2023), pp. 99–100.

R. Martins Costa Moreira, G. Wedy, *Brazilian Environmental and Climate Change Law*, https://doi.org/10.1007/978-3-031-87998-2_3

Then, in 1987, the Brundtland Report, formulated within the scope of the United Nations World Commission on Environment and Development, through a document entitled *Our Common Future*, represented a major step forward in the definition of sustainable development, stating that humanity "has the ability to make development sustainable to ensure that it meets the needs of the present without compromising the ability of future generations to meet their own needs."[2] Subsequently, Principle 4 of the Rio Declaration of 1992 stated that, in "order to achieve sustainable development, environmental protection shall constitute an integral part of the development process and cannot be considered in isolation from it."

Brundtland's description, despite its importance, has been criticized as being too anthropocentric.[3] The concept of sustainability requires a broader perspective, as it encompasses not only the fulfillment of material needs but also intangible values such as freedom, security, education, justice, and a healthy environment.[4] This concept was later expanded by Robert Solow, who introduced the idea of sustainability as the requirement to maintain for future generations the same or better standard of living that we enjoy today.[5] Amartya Sen, however, still considers this perspective insufficient, since the importance of human life lies not only in our standard of living and satisfaction of needs, but also in the freedom we enjoy. He reformulates, then, the proposals of Brundtland and Solow to include in the concept of sustainability the preservation and, where possible, the "expansion of the substantive freedoms and capabilities" of the people of today "without compromising the ability of future generations' to have similar or greater freedom."[6]

The predominant view in international,[7] European,[8] and Brazilian law is that sustainable development presents a three-dimensional scope, that is, it requires the weighted fulfillment of three pillars to achieve sustainability: ecological, social, and economic.[9] Besides that, after the formulation of the *Millennium Development Goals* during the 2012 Conference, the United Nations approved in 2015 "17 Sustainable Development Goals and 169 Targets" through the document called "Transforming our World: The 2030 Agenda for Sustainable Development."[10]

---

[2] United Nations (1987).

[3] Bosselmann (2015), p. 50.

[4] Bosselmann (2015), pp. 52–53.

[5] Solow (1993).

[6] Sen (2009).

[7] The three-pillar model of sustainability is derived, e.g., from Earth Charter, Rio Declaration 1992, Johannesburg Declaration 2002, and the United Nations Conference on Sustainable Development 2012 ("Rio+20").

[8] Gärditz and Kahl (2023) pp. 99–100.

[9] Fensterseifer and Sarlet (2023), pp. 289–296. Juarez Freitas also includes the ethical and legal-political dimensions, as intertwined aspects of the multidimensionality of well-being and sustainability (Freitas 2019, pp. 61–85). Wedy, in turn, includes, besides the environmental, economic and social pillars, the governance aspect (Wedy 2018).

[10] United Nations (2015a). Department of Economic and Social Affairs. Sustainable Development. Transforming our world: the 2030 Agenda for Sustainable Development.

This approach, however, which is due to the imprecision and openness of the concepts, as well as the conflicting goals between ecology preservation and economic interests,[11] may pose challenges to the effective implementation of environmental law and has consequently faced criticism.[12] Some scholars argue that sustainability must preserve an "ecological essence,"[13] so that environmental sustainability is understood in a stricter approach, i.e., especially through the protection and long-term safeguarding of the natural foundations of life in the sense of a lasting and environmentally compatible development.[14]

In any case, sustainable development is widely regarded as a limitation on unchecked growth, emphasizing balance and sustainability over mere economic expansion.[15] Sustainability is multidimensional and inherently tied to the development of a fraternal society,[16] as it embraces not only material aspects but also essential intangible values like freedom, security, health, education, justice, and a healthy environment. In fact, as proposed by the 2030 Agenda, sustainable development demands, in an integrated and indivisible way, the promotion of social, economic, and environmental dimensions. This holistic and integrated approach to sustainability, moreover, leads to a questioning of the GDP as the paramount development indicator of the countries, because it is based exclusively on material growth, disregarding the depreciation of natural and human resources.[17] Other indicators were developed,[18] such as the HDI (Human Development Index) and the even more advanced "*Report by the Commission on the Measurement of Economic Performance and Social Progress*" prepared by the *Stiglitz-Sen-Fitoussi Commission* in which sustainability is included as one of the factors to be assessed for an evaluation of a country's development.[19]

More than that, sustainable development is evolving toward "strong sustainability," which prioritizes preserving ecological integrity, critically reassessing economic growth as an end in itself, and steering development to respect ecosystems' carrying capacities and the finite nature of natural resources. This contrasts with

[11] Regarding the different approaches to solving the conflicting goals of sustainability, between growth and environmental protection, see: Döring (2019); Sachs (2015); Soromenho-Marques (2011).

[12] For a description of the problem in German Law: Gärditz and Kahl (2023), pp. 99–102. On the dispersive character of the sustainable development objectives, which can cover up positive points, see: da Veiga (2015), pp. 114–115.

[13] Bosselmann (2015), pp. 42–44.

[14] Gärditz and Kahl (2023), p. 100.

[15] Freitas (2019), p. 53.

[16] On the relationship between sustainability and fraternity, see: da Silva and da Silva Antunes de Souza (2020).

[17] da Veiga (2010), p. 19; Wedy (2018), position 522–552.

[18] Freitas (2019), pp. 46–47.

[19] Stiglitz et al. (2009).

"weak sustainability," which regards nature as subordinate to economic growth and places social, economic, and ecological values on an equal footing.[20]

In Brazil, legal scholars[21] and Supreme Court jurisprudence recognize sustainability as both a principle and a fundamental right derived from the Constitution. This recognition is particularly rooted in Art. 170, VI, which subordinates the Economic Order to environmental protection, and Art. 225, which declares that "all have the right to an ecologically balanced environment, which is an asset of common use and essential to a healthy quality of life, and both the government and the community shall have the duty to defend and preserve it for present and future generations."

Sustainable development can also be grounded in the constitutional primacy of human dignity,[22] which is established as a foundational value of the entire constitutional order. This principle envisions a society based not solely on liberal capitalism or quantitative economic growth but also on social and environmental justice. It seeks to promote high-standard socioeconomic and environmental development imbued with human values, transcending a simplistic, materialistic utilitarianism and the notion of welfare as merely the accumulation of capital.[23]

In conclusion, the principle of sustainability in the Brazilian legal system, while embedded in various statutes,[24] has also garnered support in both academic and practical realms as a key framework for guiding judicial review of the sustainability of administrative decisions.[25]

It is clear that defining what constitutes sustainability and the methods to achieve it in specific cases still needs further refinement. However, significant progress has been made in this direction, as in the case of sustainable or green public

---

[20] On the difference between strong and weak sustainability, see: Bosselmann (2015), p. 47; da Veiga (2015), pp. 33–34.

[21] For a defense of the fundamental right to sustainable development, it is suggested to read: Wedy (2018). Sustainability is also considered a constitutional principle of mandatory complying. For instance: Freitas (2019); Moreira (2017).

[22] Article 1, III of the Brazilian Constitution establishes the dignity of the human person as one of the foundations of the Republic of Brazil. Article 3 considers as an objective of the Republic to build a free, just, and solidary society; to guarantee national development; to eradicate poverty and substandard living conditions and to reduce social and regional inequalities; and to promote the well-being of all, without prejudice as to origin, race, sex, color, age, and any other forms of discrimination.

[23] Moreira (2017), pp. 58–59.

[24] The legal status of sustainability principle is found in a wide range legislation, such as, for instance: Atlantic Forest Act of 2006, Water Resources Act of 1997, Climate Change Act of 2009, Nature Conservation Units Act of 2000, Solid Waste Act of 2010, Forest Code of 2012, Biodiversity Protection Act of 2015, new Public Procurement Act of 2021, etc.

[25] The issue of "judicial review of administrative decisions sustainability," that is, the scope and the limits of the Judiciary reviewing actions and omissions of the government deemed to violate the sustainability principle was widely discussed in: Moreira (2017).

procurement, in which objective requirements for measuring the sustainability of public procurement were designed.[26]

Another principle that can be related to or even derived from sustainability is the principle of *intergenerational justice* or *intergenerational equity*, which assumes that all generations have an equal place in relation to the natural system, and that there is no basis for preferring past, present, or future generations.[27] It is also provided by the Brazilian Constitution, under Art. 225, which states that the protection of the environment must be ensured for both present and future generations.

3.3 Principle of the Socio-Environmental Function of Property

Modern notions of possession and ownership encompass not only the rights, but also the responsibilities and obligations entrusted to the possessor and the owner. In environmental law, it means that ecological values and rights also load the property right with a new burden of duties and obligations correlated to its exercise.[28] The social function of property finds support in the current constitutional text, either by its establishment as a fundamental right (Brazilian Constitution, Art. 5, XXIII), or by regarding it as a principle for the Economic Order (Art. 170, III).

With respect to urban property, Art. 182, § 2 of the 1988 Constitution establishes that "urban property fulfills its social function when it meets the fundamental requirements for city planning expressed in a master plan."[29] The City Statute (Law n. 10.257/01) considers, in Art. 39, that urban property fulfills its social function when it meets the fundamental requirements for the ordering of the city expressed in a master plan, ensuring that citizens' needs are met with regard to the quality of life, social justice, and the development of economic activities, respecting the guidelines set forth in Art. 2, many of which are related to the socio-environmental sustainability of cities.

For rural lands, Art. 186 of the Constitution states that social function is met when the rural property complies with, simultaneously, the criteria and standards established by legislation, the requirements of rational and proper use, sustainable use of natural resources, and preservation of the environment, compliance with the provisions that regulate labor relations and exploitation that favors the well-being of

[26] At the federal level, for example, a "Sustainable Public Procurement Guide" was published in 2020, containing more concrete parameters for the fulfillment of sustainability criteria by public biddings and procurements (Brazil 2020).

[27] Weiss (1992).

[28] Fensterseifer and Sarlet (2023), pp. 302–307.

[29] "Master Plan" or "*Plano Diretor,*" mandatory for cities with more than 20 thousand inhabitants, is a municipal statute that establishes rules, parameters, incentives, and instruments for the development of the city. It is the basic tool of urban development and expansion policy (Brazilian Constitution, Art. 182, § 1; City Statute (Law n. 10.257/2001), Art. 39 to 42-B).

the owners and workers. Likewise, Art. 6 of Law n. 8.629/93 (that deals with agrarian reform) states that productive property is considered to be that property which, when exploited economically and rationally, simultaneously reaches levels of land use and exploitation efficiency, according to indicators established by the appropriate federal agency. The relationship between fulfilling the social function and environmental protection is reinforced by Art. 9 of the same statute. Thus, the Federal Government may exercise its entitlement of compulsory purchase (eminent domain) for agrarian reform purposes due to non-compliance with the social function of possession and property (Law n. 8.629/93, Art. 2 and § 1), upon payment of compensation in the form of agrarian debt bonds (Art. 5).

The socio-environmental function of possession and property is also supported by Art. 2 of the National Environmental Policy Act (Law n. 6.938/81), which establishes several principles to be observed by possessors and owners. The Forest Protection Code (Law n. 12651/12) reiterates this provision and states that in the use and exploitation of vegetation, actions or omissions contrary to the provisions of such Act are considered unlawful use of property (Art. 2, § 1). The Forest Protection Code also prescribes the principle of governmental action for the protection and sustainable use of forests, enshrining the country's commitment to balancing productive land use with the preservation of water, soil, and vegetation (Art. 1-A, sole paragraph, III). Moreover, the environmental obligations established in this statute are regarded as obligations *in rem* and are transmitted to the successor, of any nature, in case of transfer of domain or possession of the property (Art. 2, § 2).

Finally, the Civil Code, in Art. 1228, § 1, states that the property rights must be used in accordance with its economic and social purposes and in such a way that the flora, fauna, natural beauty, ecological balance, and historic and artistic heritage are preserved, and that air and water pollution are avoided.

Therefore, by extending the long-established principle of the social function of property to also include the owner's or possessor's duty to protect the environment, it has become clear that private property cannot be exercised in an unlimited way, without respecting the ecological balance, the integrity of environmental goods, and even the rights of future generations.

3.4 Polluter-Pays Principle

The *polluter-pays principle*, in Kiss and Shelton's words, "seeks to impose the costs of environmental harm on the party responsible for the pollution."[30] Thus, the negative externalities, or indirect costs, associated with natural resource exploitation and projects that impact the environment should be considered in the regulation of economic and social activities. According to Stiglitz, externalities always arise when a certain action results in impacts on another person, for which the latter is not paid

[30] Kiss and Shelton (2007), p. 95.

or is not compensated. It brings as consequences the overproduction of goods that generate negative externalities and insufficient supply of goods that generate positive ones. This is exemplified by the contamination of air and water, which is often not effectively controlled by the actions of the polluters, as they lack sufficient incentive to invest in reducing the contamination.[31]

The existence of these negative side effects led to the formulation, in environmental law, of the *polluter-pays principle*, which requires the internalization of environmental burdens into the production process. This principle was provided for in Principle 16 of the 1992 Rio Declaration, by recommending that national authorities should seek to promote the internalization of environmental costs and the use of economic instruments, taking into account the approach that the polluter should, in principle, bear the costs of pollution with due regard to public interest and without distorting international trade and investment.

This principle, however, does not aim at tolerating pollution for a price. It cannot be distorted into the "pays to pollute" principle, that is, the entrepreneur does not pay to degrade, but to avoid environmental damage and internalize the negative externalities in the production process.[32] Its scope is even broader, including the costs of prevention, recovery, and repression of environmental damage. Associated with the *polluter-pays* principle is the *reparation principle*, which means that those who pollute, must also pay and repair for the impacts caused.[33] This standard of law, thus, tends to avoid the privatization of profits and the socialization of losses, very common in today's risk society.[34]

The legal foundation of this principle is found in Art. 4, VII, of Law n. 6.938/81, which states that the National Environmental Policy seeks to impose on polluters and exploiters the obligation to recover and/or to compensate for the damage they cause. The recognition of this principle by the Superior Court of Justice jurisprudence has been the basis for, e.g., shifting the burden of proof and requiring the person who wishes to carry out an activity to prove that it will not cause harm to the environment[35] and the imposition of liability for environmental damage, regardless of intent, negligence, or recklessness (strict liability), based on the *full risk theory*[36] (which we will analyze further in due course[37]).

Another implication of the polluter-pays principle is the *user-pays principle*, also established in Art. 4, VII, of Law n. 6.938/81, which stipulates that the National

[31] Stiglitz (2000), pp. 248–250.

[32] Milaré (2018), p. 272.

[33] Morato Leite (2015), position 3784–4002.

[34] According to Beck, the unintended side effect of the socialization of nature is the "socialization of the destruction of and threats to nature," its transformation into economic, social, and political conflicts (Beck 1998, p. 89).

[35] Superior Court of Justice, Resp. 1787748/RS, Rapporteur Min. Herman Benjamin, j. 10/12/2019.

[36] Superior Court of Justice, AgInt no AREsp 620.488/PR, Rapporteur Min. Og Fernandes, j. 04/09/2018.

[37] Section 9.2.

Environmental Policy aims to require users to contribute to the use of environmental resources for economic purposes. The *user-pays principle* is not punitive in nature; it applies irrespective of the legality of the user's actions. In essence, users of environmental resources must compensate for their use, regardless of whether they caused degradation or acted within the law, to promote responsible and sustainable consumption.

A third manifestation of this principle is the *protector-receiver* or *protector-beneficiary* principle, which, instead of repressing illegal or environmentally degrading actions, aims at encouraging and stimulating individuals and companies to adopt environmentally friendly behavior and sustainable practices through subsidies, differentiated treatment by the public administration and other benefits. It is inscribed in Art. 6, II of the Solid Waste Act of 2010 (Law n. 12.305/10) and calls for regulation by inducing and stimulating sustainable practices, usually more efficient than traditional repressive and punitive *command-and-control* measures. Article 42 of Law n. 12.305/2010, for example, stipulates the possibility of the government creating induction measures and financing lines to support ecologically sound initiatives.

More recently, based on this principle, Law n. 14.119/21 was enacted to establish the National Policy for Payment for Environmental Services. This statute explicitly emphasizes the need to adhere to this principle (Art. 5, I). Payment for environmental services functions as an incentive and promotional measure, providing monetary or non-monetary benefits (Art. 3) to individuals, legal entities (public or private), or family and community groups. These benefits are offered to those who meet eligibility criteria and actively maintain, restore, or enhance the environmental conditions of ecosystems. For a more detailed analysis of the subject, see Sect. 8.5.2.

3.5 Prevention and Precautionary Principle

The regulation of any activity or project that actually or potentially pollutes or causes environmental impact must be guided by the prevention and precautionary principles. Both, although not the same, have a similar origin in the perception that environmental damage is usually irreversible or difficult to recover. They establish the priority that should be given to measures that avoid the onset of environmental harm.[38]

The *preventive* approach has been endorsed, directly or indirectly, by the 1972 Stockholm Declaration, the 1978 UNEP Draft Principles, and the 1982 World Charter for Nature. Furthermore, this principle has been found in many treaties dealing with specific natural resources.[39]

---

[38] Milaré (2018), p. 266.

[39] Sands (2003), p. 247.

In turn, the *precautionary* principle only began to appear within international law as of the mid-1980s, although it had appeared as a principle in some domestic legal systems, most notably in West Germany. The core of the principle is reflected in Principle 15 of the Rio Declaration, which provides that: "Where there are threats of serious or irreversible damage, lack of full scientific certainty shall not be used as a reason for postponing cost-effective measures to prevent environmental degradation."[40] The precautionary principle has also been provided for in the 1992 United Nations Framework Convention on Climate Change, the 1992 Convention on Biological Diversity, and the 2012 Rio de Janeiro Declaration, among other international agreements.

Despite some controversies in Brazilian literature,[41] the prevailing understanding is that the two principles are different. Scholars generally differentiate prevention from precaution in that the former requires state action in the face of the certainty of environmental degradation, whereas for the latter, scientific uncertainty, either of the harmful effects or the causal link, is sufficient to justify the regulation, restriction, or suspension of activities. The distinction between these principles can be divided according to the estimated degree of probability of the damage occurring[42] or, as stated by Wedy, by using an imaginary causal straight line, in which the precautionary situation is prior to that in which the prevention principle applies to the hypothetical damage.[43] Another way to distinguish between the two lies in the purpose of avoiding actual danger (scientifically proven) for prevention, and of avoiding abstract danger (not scientifically proven, but likely to occur) for precaution.[44] The precautionary principle, thus, is based on the long-established idea that "it is better to be safe than sorry."

Hence, since the introduction and development of the precautionary principle under international law and within national legal systems, environmental decision-making at all levels (in the field of regulation, licensing, and planning, civil, administrative, and criminal liability) is necessarily guided not only by the certain and inexorable consequences of actions and omissions, but also by the mere probability of risk, even in the absence of scientific certainty. As Gärditz and Kahl summarize, the precautionary principle, which, together with sustainability, represents a guiding principle (*Leitprinzip*) of modern environmental policy, aims to prevent the mere probability of damage or the mere risk, through early action and cautious management of natural resources, to safeguard the ecological basis of life in

[40] Sands (2003), pp. 267–268.

[41] Departing from the predominant orientation, e.g., Fiorillo claims that the so-called "precautionary principle," if it can be observed at the constitutional level, would obviously be placed within the constitutional principle of prevention (Fiorillo 2017, pp. 91–92). For Sirvinskas, the prevention principle is a broader category in which precaution is included (Sirvinskas 2017, p. 146). Milaré, at first, adopted the prevention principle as a formula that would encompass precaution, but he went on to embrace the aforementioned difference (Milaré 2018, pp. 265–266).

[42] Freitas (2014), pp. 119–132.

[43] Wedy (2017), pp. 47–60.

[44] Wedy (2017), pp. 47–60.

the long term, subject to certain limits under the rule of law (especially following the proportionality principle).[45]

To apply the precautionary principle, the risk of damage must be plausible or there must be a well-founded conviction that damage will occur if the causal link is not timely interrupted. That is, the probability of damage must be based on reasonable grounds.[46] It cannot be the result of mere speculation, remote fear, public outcry, or ideological preference,[47] and the state's reaction to alleged threats of environmental harm cannot be driven by populism or irrational demagogy.[48]

The precautionary principle is usually analyzed from two extreme approaches: strong and weak perspectives. The first is usually associated with the demand for state action in the face of a mere abstract possibility of risk. It seeks to achieve (an unrealistic) standard of *zero risk*, with the suspension of the activity until scientific uncertainties are solved and the admission of the burden of proof reversion in class actions.[49] The minimalist or weak version of the principle requires serious and irreversible risks, rules out the standstill of activities, and does not lead to the reversal of the burden of proof.[50] This is the approach adopted by those who criticize the precautionary principle on the grounds that it is paralyzing[51] and disproportionate.

As Sunstein pointed out, in his widely quoted statement, the aggressive regulation of risks that "are unlikely to come to fruition" is an incoherent idea, "simply because regulation itself can create risks." People, he says, usually "take excessive precaution against the most available risks. And if no vivid example is associated with a particular risk, people may well be fearless – and expose themselves to real danger." Sunstein, therefore, draws attention to the problem of "probability neglect, through which intense emotions lead people to focus on worst-case scenarios without taking account of the likelihood that they will occur," and suggests that "sensible regulators manage fear through education and information," so that "cost-benefit analysis" is essential, although its result is not always decisive for decision-making.[52] He also considers that especially "if the worst-case scenario is very bad, strong precautions are entirely appropriate."[53]

Regardless of extreme versions, there is a tendency to demand proportionality and adequate justification in the precautionary measures adopted, whether they involve shutting down or regulating activities in some other way.

---

[45] Gärditz and Kahl (2023), p. 93.

[46] Freitas (2014), pp. 119–132.

[47] Along those lines, see: Freitas (2014), p. 125; Wedy (2017), pp. 62–63.

[48] Freitas (2014), p. 125.

[49] Cappelli et al. (2013), p. 56.

[50] Cappelli et al. (2013), p. 56.

[51] Sunstein (2003).

[52] Sunstein (2005), pp. 224–226. For more recent considerations by the same author, see: Sunstein (2018), position 3071–3175.

[53] Sunstein (2018), position 3071.

The Brazilian Constitution does not explicitly mention the precautionary principle. However, it can be inferred from an interpretation of the constitutional text, particularly in recognizing that the 1988 Constitution seeks to protect public health and the environment from all forms of potential harm, not just actual harm.

In environmental matters, the precautionary principle can be derived from Art. 225 of the Constitution, which guarantees everyone the right to an ecologically balanced environment. This principle is also reflected in the duties imposed on the government in § 1 of Art. 225. These duties include requiring public authorities to ensure the preparation of prior environmental impact assessments for works and activities that may potentially cause significant environmental degradation (IV) and controlling the production, sale, and use of techniques, methods, or substances that pose risks to life, quality of life, and the environment (V).

In the legislative realm, the precautionary principle is reflected in environmental law at all levels: failure to take precautionary measures in case of risk of serious and irreversible environmental damage constitutes a pollution crime (Law n. 9605/98, Art. 54, § 3) and an administrative violation (Decree n. 6.514/08, Art. 62, VII); the mere potential to cause pollution is sufficient to justify the requirement for environmental licensing (Law n. 6.938/81, Art. 8, I; Art. 9, IV), land control and zoning (Art. 2, V); the management, disposal, and destination of solid waste must avoid risks to public health and safety, supported by the precautionary principle (Law n. 12.305/10, Art. 3, VII e VIII; Art. 6, I); in the protection and use of the Atlantic Forest biome the precautionary principle must be observed (Law n. 11.428/06, Art. 6, sole paragraph); the National Policy on Climate Change and the actions resulting therefrom, carried out under the responsibility of political entities and public administration bodies, must observe the principles of precaution and prevention, and measures must be taken to foresee, avoid, or minimize the identified causes of climate change of anthropogenic origin inside the national territory (Law n. 12.187/09, Art. 3, II); the handling of genetically modified organisms must observe the precautionary principle (Law n. 11.105/05, Art. 1).

Brazilian courts increasingly rely on the precautionary principle as a foundation for their decisions. The Superior Court of Justice uses this principle to support the adoption of the *full risk theory* (*teoria do risco integral*) in cases of civil liability for environmental damage, effectively rejecting the degrader's claims of exoneration, as we will discuss in due course.[54] The Court also applies the precautionary principle to shift the burden of proof in environmental lawsuits, requiring the alleged degrader to demonstrate that their activities do not cause environmental harm.[55] Additionally, the Superior Court has invoked this principle to suspend licenses until uncertainties

[54] Section 9.2.

[55] This understanding is portrayed in the Summula (restatement of case law) ("*súmula*") n. 618 of Superior Court of Justice, in these terms: "the reversion of the burden of proof applies to environmental lawsuits."

about a project's impacts are resolved[56] and to require environmental licensing and environmental impact assessment before authorizing the burning of sugarcane straw.[57]

Similarly, the Supreme Court has recognized and applied the precautionary principle, e.g., to: (a) justify the ban on importing used and remolded tires;[58] (b) suspend a mining activity due to evidence of serious damage to the environment;[59] (c) require prior approval by health and environmental authorities and proof of effectiveness for the use of insecticide dispersal by aircraft to combat the *aedes aegypti* mosquito;[60] (d) quash the validity of environmental deregulation measures, such as the attempt to revoke CONAMA's[61] regulations, without replacing or updating them;[62] (e) decide for the unconstitutionality of the suspension by the Executive branch of the *closed season* (*período de defeso*) for fishing in a certain period;[63] (f) deem unconstitutional a state rule that establishes hypotheses for dispensing with and simplifying environmental licensing for open-pit mining activities.[64]

However, in another decision involving electromagnetic field control by electric power utilities, the Supreme Court held that the precautionary principle cannot generate paralysis, nor can it justify measures based on unfounded fear, and that it is necessary to observe the proportionality of the measures to be adopted.[65]

3.6 Principle of Publicity, Transparency, or Access to Information in Environmental Matters

The principle of publicity, or transparency, is a cornerstone of public law and should, as a rule, guide all activities of public authorities. Environmental law is no exception: the imperative for transparency in information and decision-making in this field is strongly upheld. The rationale is clear: environmental balance is regarded as a *public*

---

[56] The case involved the expansion of a coastal avenue: Superior Court of Justice, AgRg na SLS 1.524/MA, Rapporteur Min. Ari Pargendler, j. 02 Mai 2012.

[57] Superior Court of Justice, AgInt no AREsp 1071566/SP, j. 23 Apr. 2019.

[58] Brazilian Supreme Federal Court, ADPF 101, j. 24 Jun. 2009.

[59] Brazilian Supreme Federal Court, SL 933 ED/PA, j. 31 Mai. 2017.

[60] Brazilian Supreme Federal Court, ADIn 5592, j. 11 Sept. 2019.

[61] CONAMA is the short for "*Conselho Nacional do Meio Ambiente*" (National Council for the Environment), which is a federal body responsible for promulgating environmental rules, among other functions. We will study CONAMA on Sect. 7.2.

[62] Brazilian Supreme Federal Court, ADPF 747, ADPF 748, ADPF 749, Rapporteur Min. Rosa Weber, j. 30 Nov. 2020.

[63] Brazilian Supreme Federal Court, ADIn 5447, Rapporteur Min. Roberto Barroso, j. 22 Mai 2020.

[64] Brazilian Supreme Federal Court, ADIn 6650/SC, Rapporteur Min. Carmen Lúcia, j. 26 Apr. 2021.

[65] Brazilian Supreme Federal Court, RE 627189/SP, Rapporteur Min. Dias Toffoli, j. 8 Jun. 2016.

good in the economic sense, and environmental assets are considered *common resources*.[66] In addition, the complexity of environmental issues, the trade secrets of many economic operations, and the traditional opacity of the public administration itself may result in *asymmetric information*[67] between, on the one hand, the entrepreneurs and the public authority, holders of knowledge and data related to a certain economic activity, and, on the other, the entire society, which is often unaware of the seriousness of the damage that is being caused to its health and environmental balance. In this regard, as Beck pointed out, many of the new risks (nuclear or chemical contamination, harmful substances in food, civilization's diseases) are being completely removed from immediate human perception.[68]

The necessity for transparency in environmental information and decision-making is widely acknowledged on the international stage, significantly influenced by the foundational provisions of the 1972 Stockholm Declaration and the 1992 Rio Declaration, particularly Principle 10 of Rio 92, which contemplates three essential aspects for dealing with environmental issues: access to information, participation in decision-making, and access to justice. These dimensions are developed, in the European context, by the Aarhus Convention of 1998, and, in Latin America, by the Escazú Agreement of 2018. In addition, one of the targets of the UN Sustainable Development Goal (SDG) n. 16 is: "Ensure public access to information and protect fundamental freedoms, in accordance with national legislation and international agreements."[69]

In the Brazilian legal system, the right to broad access to environmental information and knowledge is enshrined in both the Constitution and legislation. However, the effective implementation and consolidation of environmental transparency remain areas requiring further development.

Some constitutional provisions are worth noting. Article 225, § 1, IV of the Brazilian Constitution prescribes that publicity must be given to environmental impact assessments. Article 225, § 1, VI states that the government must "promote environment education in all school levels and public awareness of the need to preserve the environment." Moreover, a brief overview of the Brazilian environmental legislation reveals the dissemination and consolidation of the transparency principle, although, in practice, access to environmental information is still inadequate. Along these lines, we can mention, e.g.: a) the mandatory publicity of the licensing process and the environmental impact assessment, safeguarding industrial confidentiality (CONAMA Resolution[70] n. 237/97, Arts. 3, 10, II and VIII; and

---

[66] Regarding the legal framework of environmental goods in Brazilian law, as *public goods* and *common resources*, see below in Sect. 4.4.

[67] On *asymmetric information* or *inadequate information* as a market failure calling for the need for administrative regulation, see: Breyer (2011), pp. 6–7.

[68] Beck (1998), p. 33.

[69] United Nations (2015b). Department of Economic and Social Affairs. *Sustainable Development.*

[70] As mentioned above, CONAMA is short for *National Council for the Environment.* CONAMA has the authority to issue environmental regulations called *Resolutions* (*Resoluções*).

Resolution n. 01/86, Art. 11); b) the provision by Law n. 6.938/81, that the development of mechanisms to promote environmental information is a regulatory tool of the National Environmental Policy (Art. 9, VII and XI). This statute also provides that the National Environmental Policy will steer the diffusion of environmental management technologies, the dissemination of environmental data and information, and the development of public awareness about the need to preserve environmental quality and ecological balance (Art. 4, V); c) the provision by Law n. 10.650/03 for public access to environmental information; d) the need for proper labeling of pesticides (Law n. 7.802/89, Art. 7, inc. I to IV and § 1); e) the labeling of transgenic products (Law n. 11.105/05, Art. 40); f) the imposition of prior informed consent granted by indigenous population or traditional community for access to traditional knowledge associated with genetic heritage, of identifiable origin (Law n. 13.123/15, Art. 9); g) the provision for public access to the data tracking system of forest products (Law n. 12651/12, Art. 35, § 4) and for the seizure of properties due to illegal devastation (Art. 51, § 2), as well as the creation of the *Rural Environmental Registry* (CAR or *Cadastro Ambiental Rural*), as part of the National Environmental Information System, a national electronic public registry, mandatory for all rural properties, with the purpose of integrating environmental information from rural properties and possessions, composing a database for the control, monitoring, environmental and economic planning, and the prevention of deforestation (Art. 29).

A recent Superior Court of Justice decision shed light on environmental transparency and established guidelines for access to environmental information. The Court stated that the right of access to information in Brazilian environmental law comprises: (a) the duty to publish, on the Internet, environmental documents held by the Administration not subject to secrecy (active transparency); (b) the right of any person and legal entity to request access to specific unpublished environmental information (passive transparency); and (c) the right to request the production of environmental information not available to the Administration (reactive transparency). It was also emphasized that the state's obligation to ensure environmental transparency is presumed, and it is the public administration's burden to justify its non-compliance, always subject to judicial review.[71]

3.7 Principle of Public Participation

Public participation is based on the right of those who may be affected to have a say in the determination of their environmental future.[72] Not, however, mere formal participation, or through elections only, but effective participation, in the context of a

[71] Superior Court of Justice, REsp 1.857.098/MS, Rapporteur Min. Og Fernandes, j. 11 Mai 2022.
[72] Kiss and Shelton (2007), p. 102.

substantial democracy[73] or, in Rosanvallon's expression, "exercise democracy," where citizens cease to be "one-day sovereigns" and participate in a more constant way in the control of the government.[74] As Rousseau wrote, democracy is an unfinished and continuous regime, which does not end with elections, nor does it stop at state borders, but is implemented between electoral moments and opens up to the space-world (*s'ouvre sur l'espace-monde*).[75]

Several international agreements since the Stockholm Conference call upon states to take specific measures to ensure that the public is adequately informed about the risks posed to them by specific activities,[76] such as the Convention on the Transboundary Effects of Industrial Accidents (Helsinki, 1992), the Convention on Environmental Impact Assessment in a Transboundary Context (Espoo, 1991), and the Cartagena Protocol on Biosafety (Montreal, 2000) to the Convention on Biological Diversity.

It is worth mentioning Principle 10 of Rio 92 Declaration, which states that: "a) environmental issues are best handled with the participation of all concerned citizens, at the relevant level"; b) at the national level, each individual shall have "the opportunity to participate in decision-making processes"; and c) "States shall facilitate and encourage public awareness and participation by making information widely available."

This principle was developed by the 1998 Aarhus Convention and, at the Latin American level, by the Escazú Agreement of 2018. Likewise, among the targets of SDG 16 of the 2030 Agenda, one can mention the need to "ensure responsive, inclusive, participatory and representative decision-making at all levels" and "broaden and strengthen the participation of developing countries in the institutions of global governance."[77]

At the Brazilian constitutional level, the sole paragraph of Art. 1 of the Constitution states that power emanates from the people and can be exercised directly, which adds the dimension of participatory democracy, opening space for the direct intervention of Brazilian citizens in public decisions.[78] For environmental issues, Art. 225 of the Constitution declares the ecologically balanced environment to be an "asset of common use and essential to a healthy quality of life" and imposes its defense and preservation not only on the government, but also on every person. Besides, in § 1, VI of Art. 225, it is prescribed as a duty of the state "to promote

---

[73] In Moreira Neto's lesson, democracy can no longer be considered just a formal process of choosing who should govern us, but also a choice of how we want to be governed (de Figueiredo Moreira Neto 2007, pp. 343–345), because citizens do not lose their freedom with the expression of their vote (Sartori 1994, p. 56).

[74] Rosanvallon (2015), p. 348.

[75] Rousseau (2015), position 122.

[76] Kiss and Shelton (2007), p. 103.

[77] United Nations (2015b). Department of Economic and Social Affairs. Sustainable Development.

[78] Fensterseifer and Sarlet (2023), pp. 312–317.

environmental education in all school levels and public awareness of the need to preserve the environment."

As Kiss and Shelton put it, "participation may take place through elections, grass roots action, lobbying, public speaking, hearings, and other forms of governance, whereby various interests and communities participate in shaping the laws and decisions that affect them."[79]

The Brazilian legislation provides several of these participative mechanisms in environmental issues, such as:

(a) the requirement to conduct public hearings before licensing projects and activities that may cause significant environmental degradation, in which an environmental impact study will also be necessary;[80]
(b) the introduction of a public civil suit[81] and people's legal suit[82] to seek liability for damage caused to the environment and to goods and rights of artistic, aesthetic, historical, touristic, and scenic value;
(c) the possibility for any person, upon finding an environmental violation, to report it to the public authorities, for the purpose of exercising their monitoring and inspection power;[83]
(d) the participation of civil society, local communities, non-governmental organizations, private entities, and individuals in the National System of Nature Conservation Units;[84]
(e) the adoption of the principle of *democratic management* in the protection and use of the Atlantic Forest Biome;[85]
(f) the determination of observance, in the National Policy on Climate Change, of the principle of *citizen participation*;[86]
(g) the assumption, as principles of the National Policy of Solid Waste, of cooperation among the different spheres of government, the business sector, and other segments of society, and of the right of society to information and social control;[87]
(h) the social participation in public boards or councils with responsibility for environmental regulation, being CONAMA the most relevant case, which is

[79] Kiss and Shelton (2007), p. 103.

[80] Brazilian Constitution, Art. 225, § 1, inc. IV; CONAMA Resolution n. 01/1986 and 09/1987.

[81] Brazilian Constitution, Art. 129, inc. III and § 1; Law n. 7.347/85.

[82] The Constitution grants standing to any citizen to propose a people's legal suit (*ação popular*) to repeal an act that is harmful to the environment and to the historical and cultural heritage (CF, Art. 5, LXXIII). The people's legal suit is governed by Law n. 4.717/65.

[83] Law n. 9.605/98, Art. 70, § 2.

[84] Law n. 9.985/2000, Art. 5, II, III, IV and V.

[85] Law n. 11.428/2006, Art. 6.

[86] Law n. 12.187/2009, Art. 3.

[87] Law n. 12.305/2010, Art. 6, V and X.

composed of members of the federal, state, and municipal governments, economic and business sectors, and civil society.[88]

The participation of civil society in the composition of CONAMA was significantly impaired following the promulgation of the Decree n. 9.806/2019. This Decree was challenged by a judicial review lawsuit before the Supreme Court,[89] which struck down the Act, deeming it at odds with the Constitution. As the Court has reasoned, these changes have prevented real opportunities for social participation in the public body, causing an irrecoverable democratic deficit. The judges highlighted that CONAMA also has a deliberative function and is a public forum for the creation of broad and sectorial environmental policies. This situation shows that, under the changes introduced by this Decree, representatives of civil society would not have a genuine opportunity to influence decision-making, being relegated to the isolated position of a minority in expressing their interests within the collective will.

Nevertheless, more recently, following the change in the presidency in 2023, Decree n. 11.417/23 was promulgated, repealing Decree n. 9.806/2019 and reinforcing social participation in CONAMA.

Brazilian Supreme Court has already spoken upon the unconstitutionality of reducing the participation of civil society in three other environmental bodies: the Deliberative Council of the National Environmental Fund, the National Council of Legal Amazon, and the Guidance Committee of the Amazon Fund. In this case, the Court found unconstitutional rules that, under the pretext of restructuring environmental agencies, remove the participation of civil society and state governors from the development and formulation of public policies.[90]

Despite the extensive legal framework for public participation in environmental decision-making, there remains a significant gap in mechanisms that offer meaningful opportunities to truly influence deliberations on environmentally significant activities. Formal or superficial participation—reduced to a mere bureaucratic step to legitimize environmental impacts, as exemplified by the "inform, invite, and ignore"[91] formula—is insufficient. This issue is particularly evident in public hearings during licensing processes, where community involvement is often inadequate and occurs too late, with contributions from individuals and associations failing to be genuinely considered.[92]

[88] The composition of CONAMA is stipulated in Art. 5 of Dec. 99.274/1990.

[89] Brazilian Supreme Federal Court, ADPF 623, Rapporteur Min. Rosa Weber, j. 22 Mai 2023.

[90] Brazilian Supreme Federal Court, ADPF 651/DF, Rapporteur Min. Cármen Lúcia, j. 28 Apr. 2022.

[91] Clarke and Peterson (2016), pp. 31–38.

[92] The limitations of public hearing as a mechanism to promote the effective participation of the population affected by a project and the consideration of their perspectives are widely studied. On the shortcomings related to late, unrepresentative, and authoritarian public hearings in Brazil, held after a decision has been made, without prior discussion of its feasibility and without stimulating the participation of the affected community, see: Vitorelli (2022), pp. 599–602. Likewise, regarding the insufficiency of public hearings to guarantee a productive dialogue and solutions that contemplate the interests of all parties, held only to comply with legal requirements, including criticism in North American Law, check out: de Souza (2014), pp. 79–80.

3.8 Principle of Non-Regression

The principle of non-regression has, in Collins's words, "both ecological and legal dimensions: it can be defined as a prohibition on state conduct that results in environmental degradation (e.g., increased pollution, biodiversity loss) or in the weakening of environmental laws."[93] Depending on the jurisdiction, non-regression may also be known as "non-degradation," "the standstill doctrine," or the obligation not to "backtrack" from environmental commitments.[94]

The principle of non-regression originates from International Human Rights Law, meaning that social and economic rights, once attained, simultaneously become both an institutional obligation and an individual entitlement.[95] Indeed, it has taken its roots from the international protection of economic and social rights, which requires the progressive and effective implementation of social protection by states, as can be noticed in the clause or duty of progressiveness set forth under the 1966 International Covenant on Economic, Social and Cultural Rights (ICESCR), and upon the duty of progressive development under Art. 26 of the 1969 American Convention on Human Rights, complemented by the Art. 1 of the 1988 Protocol of San Salvador Additional to the American Convention on Human Rights in the Area of Economic, Social and Cultural Rights.[96]

The prohibition of retrocession is closely related to the notion of stability and predictability of the legal system, as an unavoidable expression of the rule of law.[97] Therefore, the doctrine of constitutional law has recognized, despite the ongoing controversies, that the achievements attained by society are safe from regressive legislative measures; every other legislation that tries to promote social regression shall be deemed unconstitutional. Here it is sought to broaden the spectrum of this principle to reach the prohibition of laws to diminish environmental protection.

The environmental policy lies precisely in the search for a better human and animal welfare, a permanent progress of society, which is why it is concluded that disproportionate regressions should be prohibited.[98] In other words, public authorities are not allowed to promote deconstructions or setbacks of the levels of environmental protection already achieved, especially to follow the constitutional obligation exactly in the opposite direction, i.e., that the government must ensure the progressive effectiveness of the right to a balanced environment, as outlined in Art. 225, § 1 of the Constitution.[99]

---

[93] Collins (2021), p. 205.

[94] Collins (2021), p. 205.

[95] Canotilho (1993), pp. 468–469.

[96] On this subject: Fensterseifer and Sarlet (2023), pp. 474–485.

[97] Sarlet (2015), pp. 464–466.

[98] Prieur (2011), p. 88.

[99] Moreira and Wedy (2019), p. 84.

This principle applies to preventing both legal and factual setbacks, as well as prohibiting substantial and procedural or organizational retrogressions. In the latter case, the principle would prevent a disproportionate downsizing or the extinction of the administrative structure currently in place in Brazil to give effect to the fundamental right to a healthy environment and to enforce environmental law.[100] Under Brazilian legislation, non-regression is mainly related to the principle of constant improvement of environmental protection, prescribed in Art. 2 of the Law n. 6.938/81.[101]

Nevertheless, such a principle cannot immobilize legislative and administrative action, so it cannot prevent, especially in times of economic crisis, adjustments, restrictions, and even normative and structural reductions. Accordingly, as Fensterseifer and Sarlet stated, the principle does not refrain from all restrictions against ecological rights, but the measures that reduce the level of protection might be considered unlawful, requiring sound justification from the state and stricter control from the Judiciary and other authorities.[102] The application of the non-regression principle, then, demands the use of the proportionality principle, to prevent excessive measures or insufficient protection.

Brazilian courts have gradually defined the contours of the principle of non-regression. In a case of great relevance, the Supreme Court, when awarding judgment of several constitutional lawsuits brought against provisions of the New Forest Protection Act (Law n. 12.651/12, which will be analyzed more closely below[103]), gave a more lenient interpretation to this principle, understanding that it is not appropriate to disqualify a particular legal rule as contrary to the constitutional command of environmental protection (CF, Art. 225), or even under the generic and subjective label of "environmental regression," ignoring the various issues that permeate the legislator's decision-making process, democratically vested with the function of appeasing conflicting interests by means of general and objective rules. Furthermore, the Supreme Court has established that the non-regression principle neither overrides the democratic principle, in the desire to transfer to the Judiciary, functions inherent to the Legislative and the Executive branches, nor justifies the removal of more efficient legal arrangements for the sustainable development of the country as a whole.[104]

However, the Supreme Court has applied this principle to prevent the essential core of the right to the environment from being harmed. Thus, in another decision, in which it blocked an attempt to reduce or suppress specially protected territorial

[100] Fensterseifer and Sarlet (2023), pp. 487–489.

[101] Article 2 of Law n. 6.983/81 states that: "The National Environmental Policy has as its objective the *preservation, improvement and recovery of the environmental quality favorable to life*, aiming to ensure, in the country, conditions for socio-economic development, the interests of national security and the protection of the dignity of human life."

[102] Fensterseifer and Sarlet (2023), pp. 474–485.

[103] Section 12.3.

[104] Brazilian Supreme Federal Court, ADC 42/DF, ADI 4901/DF, ADI 4902/DF, ADI 4903/DF e ADI 4937/DF, Rapporteur Min. Luiz Fux, j. 28 Feb. 2018.

spaces by means of a *provisional measure* (*medida provisória*)[105] adopted by the President of the Republic, the Supreme Court held that, while the application of the non-regression principle may not freeze legislative and administrative actions, because it is necessary to admit a certain margin of discretion for public authorities in environmental matters, the *essential core* (*núcleo essencial*) of the fundamental right to an ecologically balanced environment should not be impaired.[106] The same Court has also considered unconstitutional, due to the violation to the non-regression principle, state legislation that exempts licensing for activities with environmental impacts.[107]

The Superior Court of Justice has also upheld the principles of non-regression and the improvement of environmental quality. For example, it has confirmed the prohibition of construction in permanent preservation areas, except in the scenarios expressly allowed by the Forest Protection Act, even in regions that are already occupied, ecologically degraded, or damaged.[108]

3.9 Principle of *In Dubio Pro Natura*

The vagueness and imprecision that characterize much of environmental legislation, combined with the scientific uncertainty surrounding the causal interaction of many activities with environmental impacts, demand careful legal interpretation to resolve challenging and complex cases. One guiding standard is the principle of *in dubio pro natura* (*in doubt, favor nature*), which, as Benjamin observed, directs judges to select the interpretation that best ensures the preservation of essential ecological processes and biodiversity when multiple meanings are possible.[109]

The application of this principle cannot serve as an easy solution for addressing complex problems, nor does it imply that the most environmentally favorable measure will always prevail. In a case concerning the burning of sugarcane straw, the Supreme Court ruled that the legislation of the state of São Paulo, which provided for the gradual elimination of this activity, took precedence over the legislation of the municipality of Paulínia, which completely prohibited burning within its territory. The Court held that this matter involved a regional interest rather

[105] The *provisional measure* (*medida provisória*) is a law adopted by the President of the Republic in important and urgent cases, which shall be submitted to the National Congress immediately to be ratified, and thereby become final law, or rejected. It is governed by Article 62 of the Brazilian Constitution.

[106] Brazilian Supreme Federal Court, ADI 4717/DF, Rapporteur Min. Cármen Lúcia, j. 05 Apr. 2018.

[107] Brazilian Supreme Federal Court, ADI 6288, Rapporteur Min. Rosa Weber, j. 23 Nov. 2020.

[108] Superior Court of Justice, REsp 769.753/SC, Rapporteur Min. Herman Benjamin, j. 08 Sept. 2009.

[109] Benjamin (2014).

than merely a local one, rejecting the argument that the most environmentally favorable regulation must always prevail.[110]

More recently, however, the Supreme Court deemed constitutional a state law prohibiting the use of animals for the development, experimentation, and testing of cosmetic products, personal hygiene items, perfumes, and their components. The Court reasoned that states are empowered to enact more protective environmental regulations based on their specific circumstances.[111]

This principle is upheld by the jurisprudence of the Superior Court of Justice, which states legislation protecting vulnerable communities and collective interests must be interpreted in the most favorable way for them, effectively facilitating access to justice and the underlying purpose of the law.[112]

Thus, the superior court has applied the principle, for instance, to rule that the absence or impossibility of technical evidence does not preclude the recognition of environmental damage and the subsequent duty of full remediation in cases of contamination caused by clandestine and illegal sewage discharge.[113] In another case, involving a real estate project potentially harmful to the environment, the court held that in situations involving impacts on a biome with exceptionally rich and endangered biodiversity, the *in dubio pro natura* principle prevails, giving rise to two main effects: *First*, the burden of proof is reversed, requiring the developer to demonstrate the absence of harm to the specially protected ecosystem. *Second*, in terms of the interpretation of legal provisions and evaluation of evidence, the judge is obligated, in cases of doubt, to decide in favor of the threatened environment. This is based on an abstract assessment of potential risk rather than a concrete assessment of actual damage.[114]

References

Beck U (1998) Weltrisikogesellschaft. Suhrkamp, Frankfurt am Main

Benjamin AHV (2014) Hermenêutica do Novo Código Florestal. Revista de Direito Ambiental 73: 15–24

Bosselmann K (2015) Princípio da Sustentabilidade: transformando direito e governança. Trans. Phillip Gil França. Revista dos Tribunais, São Paulo. English edition: Bosselmann K (2008) The principle of sustainability: transforming law and governance. Ashgate, Aldershot

---

[110] Brazilian Supreme Federal Court, RE 586.224/SP, j. 05 Mar. 2015.

[111] Brazilian Supreme Federal Court, ADI 5.996, j. 15 Apr. 2020.

[112] Superior Court of Justice, REsp 1198727/MG, Rapporteur Min. Herman Benjamin, j. 14 Aug. 2012; REsp 1255127/MG, Rapporteur Min. Herman Benjamin, j. 18 Aug. 2016; REsp 1356207/SP, Rapporteur Min. Paulo de Tarso Sanseverino, j. 28 Apr. 2015; REsp 1462208/SC, Rapporteur Min. Humberto Martins, j, 11 Nov. 2014.

[113] Superior Court of Justice, REsp No. 2.065.347/PE, Rapporteur Min. Francisco Falcão, j. 27 Feb. 2024.

[114] Superior Court of Justice, REsp n. 1.860.232/SC, Rapporteur Min. Francisco Falcão, j. 23 Mai. 2023.

Brazil (2020) Advocacia-Geral da União/Consultoria-Geral da União. Guia Nacional de Contratações Sustentáveis. file:///C:/Users/bch/Downloads/guia_nacional_de_contratacoes_sustentaveis_-_3__edicao_abr_2020.pdf. Accessed 18 Nov 2024

Breyer S et al (2011) Administrative law and regulatory policy: problems, text, and cases, 7th edn. Wolters Kluwer Law & Business, New York

Canotilho JJG (1993) Direito Constitucional, 6th edn. Almedina, Coimbra

Cappelli S, Marchesan AMM, Steigleder AM (2013) Direito Ambiental, 7th edn. Verbo Jurídico, Porto Alegre

Clarke TL, Peterson TR (2016) Environmental conflict management. SAGE Publications, London

Collins L (2021) Principle of Non-Regression. In: Morin J-F, Orsini A (eds) Essential concepts of global environmental governance, 2nd edn. Routledge, New York, pp 205–206

da Silva IRV, da Silva Antunes de Souza MC (2020) Fraternidade e Sustentabilidade: diálogo necessário para a formação de uma consciência ecológica e construção de uma sociedade fraterna. In: da Silva Antunes de Souza MC (coord) Direito e Sustentabilidade: temas contemporâneos. Lumen Juris, Rio de Janeiro, pp 313–332

da Veiga JE (2010) Sustentabilidade: a legitimação de um novo valor, 2nd edn. SENAC, São Paulo

da Veiga JE (2015) Para entender o desenvolvimento sustentável. Editora 34, São Paulo

de Figueiredo Moreira Neto D (2007) Mutações do Direito Administrativo, 3rd edn. Renovar, Rio de Janeiro

de Souza LM (2014) Resolução consensual de conflitos envolvendo políticas públicas. Fundação Universidade de Brasília, Brasília

Döring T (2019) Alternativen zum umweltschädlichen Wachstum. Wirtschaftsdienst 99(7): 497–504

Fensterseifer T, Sarlet IW (2023) Princípios do Direito Ambiental. Saraiva, São Paulo

Fiorillo CAP (2017) Curso de Direito Ambiental Brasileiro, 17th edn. Saraiva, São Paulo

Freitas J (2014) Direito Fundamental à Boa Administração Pública, 3rd edn. Malheiros, São Paulo

Freitas J (2019) Sustentabilidade: Direito ao Futuro, 4th edn. Forum, Belo Horizonte

Gärditz KF, Kahl W (2023) Umweltrecht, 13th edn. C.H. Beck, München

Kiss A, Shelton D (2007) Guide to international environmental law. Martinus Nijhoff, Leiden

Milaré É (2018) Direito do Ambiente, 11th edn. Revista dos Tribunais, São Paulo

Morato Leite JR (2015) Sociedade de Risco e Estado. In: Canotilho JJG, Morato Leite JR (eds) Direito Constitucional Ambiental Brasileiro, 6th edn. Saraiva, São Paulo. Kindle Edition

Moreira RMC (2017) Direito Administrativo e Sustentabilidade: o novo controle judicial da Administração Pública. Forum, Belo Horizonte

Moreira RMC, Wedy G (2019) Manual de Direito Ambiental: de acordo com a jurisprudência dos Tribunais Superiores. Forum, Belo Horizonte

Prieur M (2011) Droit de l'environnement, 6th edn. Dalloz, Paris

Rosanvallon P (2015) El buen gobierno. Trad. Horacio Pons. Manantial, Buenos Aires

Rousseau D (2015) Radicaliser la démocracie: propositions pour une refondation. Seuil, Paris. Kindle Edition

Sachs J (2015) The age of sustainable development. Columbia University Press, New York

Sands P (2003) Principles of international environmental law, 2nd edn. Cambridge University Press

Sarlet IW (2015) A Eficácia dos Direitos Fundamentais: uma teoria geral dos direitos fundamentais na perspectiva constitucional, 12th edn. Livraria do Advogado, Porto Alegre

Sartori G (1994) A teoria da democracia revisitada. V. 1. O debate contemporâneo. Trad.: Dinah de Abreu Azevedo. Ática, São Paulo

Sen A (2009) The idea of justice. Belknap Press of Harvard University Press, Cambridge

Sirvinskas LP (2017) Manual de Direito Ambiental, 15th edn. Saraiva, São Paulo

Solow RM (1993) An almost practical step toward sustainability. Resources Policy 19(3):162–172

Soromenho-Marques V (2011) Desenvolvimento sustentável: correntes e polémicas em tempos difíceis. Revista Dirigir 115–116:3–9

Stiglitz J (2000) La Economía del Sector Público. Transl. Maria Esther Rabasco e Luis Toharia, 3rd edn. Antoni Bosch, Barcelona. English edition: Stiglitz J (2000) 3rd edn. WW Norton & Company, New York

Stiglitz JE, Sen A, Fitoussi J-P (2009) Report by the Commission on the Measurement of Economic Performance and Social Progress. Commission on the Measurement of Economic Performance and Social Progress. http://www.insee.fr/fr/publications-et-services/dossiers_web/stiglitz/doc-commission/RAPPORT_anglais.pdf. Accessed 18 Nov 2024

Sunstein C (2003) The paralyzing principle. Regulation 25(4):32–37

Sunstein C (2005) Laws of fear: beyond the precautionary principle. Cambridge University Press, New York

Sunstein CR (2018) The cost-benefit revolution. The Mit Press, Cambridge. Kindle edition

United Nations (1987) Department of Economic and Social Affairs. Sustainable Development. Report of the World Commission on Environment and Development: Our Common Future. https://sustainabledevelopment.un.org/content/documents/5987our-common-future.pdf. Accessed 18 Nov 2024

United Nations (2015a) Department of Economic and Social Affairs. Sustainable Development. Transforming our world: the 2030 Agenda for Sustainable Development. https://sdgs.un.org/2030agenda. Accessed 18 Nov 2024

United Nations (2015b) Department of Economic and Social Affairs. Sustainable Development. https://sdgs.un.org/goals/goal16. https://sdgs.un.org/topics/information-integrated-decision-making-and-participation. Accessed 18 Nov 2024

Vitorelli E (2022) O devido processo legal coletivo: dos direitos aos litígios coletivos, 3rd edn. Revista dos Tribunais, São Paulo

Wedy G (2017) O Princípio Constitucional da Precaução: como instrumento de tutela do meio ambiente e da saúde pública, 2nd edn. Fórum, Belo Horizonte

Wedy G (2018) Desenvolvimento Sustentável na Era das Mudanças Climáticas: um direito fundamental. Saraiva, São Paulo. Kindle Edition

Weiss EB (1992) In fairness to future generations and sustainable development. Am Univ Int Law Rev 8:19–26

Chapter 4
Environmental Constitutionalism in Brazil

4.1 Introduction

In parallel to the right to life, liberty, equality, health, or human dignity itself, from the second quarter of the twentieth century on, the right to a healthy environment has experienced a remarkable development in international law, in such a way as to influence several legal systems and national constitutions to contemplate environmental protection and regulations, although at different levels of detail and normative density.

According to the First Global Report on the Environmental Rule of Law, published in 2019 by UNEP (UN Environment Programme), since the 1970s, 88 countries have adopted a constitutional right to a healthy environment, with an additional 62 countries enshrining environmental protection in their constitutions in some form—a total of 150 countries from all over the globe with constitutional rights and/or provisions on the environment.[1]

Brazil has followed a similar path of progressive development of an environmental regulatory framework, with the 1988 Constitution marking a milestone in environmental protection. Indeed, it was in the current Constitution that environmental protection was granted the status of a constitutional right, including a chapter specifically dedicated to the environment (Art. 225).[2]

[1] United Nations Environment Programme (2019).
[2] da Silva (2010), p. 48; Thomé (2017), pp. 115–116.

R. Martins Costa Moreira, G. Wedy, *Brazilian Environmental and Climate Change Law*, https://doi.org/10.1007/978-3-031-87998-2_4

4.2 Environmental Law in the Brazilian Constitution

The human-environment relationship is regulated under Art. 225 of the Constitution and its paragraphs, and the *caput* of Art. 225 states that "all have the right to an ecologically balanced environment, which is an asset of common use and essential to a healthy quality of life, and both the government and the community shall have the duty to defend and preserve it for present and future generations." Section 1 of Art. 225 lists several duties imposed on the government regarding environmental protection, each of which will be analyzed in due course. Article 225, § 2 establishes that: "those who exploit mineral resources shall be required to restore the degraded environment, in accordance with the technical solutions demanded by the competent public agency, as provided by law." Article 225, § 3 states that: "Procedures and activities considered as harmful to the environment shall subject the infractors, be they individuals or legal entities, to criminal and administrative sanctions, without prejudice to the obligation to repair the damages caused." As Art. 225, § 4 prescribes: "The Brazilian Amazonian Forest, the Atlantic Forest, the Serra do Mar, the Pantanal Mato-Grossense and the coastal zone are part of the national patrimony, and they shall be used, as provided by law, under conditions which ensure the preservation of the environment, therein included the use of mineral resources." For Art. 225, § 5: "The unoccupied lands or lands seized by the states through discriminatory actions which are necessary to protect the natural ecosystems are inalienable." Section 6 states that: "Power plants operated by nuclear reactor shall have their location defined in federal law and may not otherwise be installed." Finally, the recently added § 7[3] establishes that sporting practices involving animals are not considered cruel, as long as they are cultural expressions, in accordance with § 1 of Art. 215 of the Constitution. To avoid being considered cruel to animals, these practices must be registered as intangible assets of Brazilian cultural heritage and regulated by specific laws that ensure the welfare of the animals involved. Regarding this last constitutional provision, it deserves further analysis, which shall be done later.[4]

The most recent constitutional change concerning the environment was introduced through Constitutional Amendment n. 123/2022, which was subsequently modified by Amendment n. 132/2023.[5] Through this amendment, item VIII was included (and changed in 2023) in § 1 of Art. 255, to provide as the government's duty, in short, to maintain a favorable tax regime for biofuels and low-carbon hydrogen, in the form of Supplementary Law, to ensure a tax rate lower than that applied to fossil fuels, thereby ensuring a competitive advantage over them.

---

[3]Constitutional Amendment n. 96/17 included § 7 to Art. 225 of the Brazilian Constitution, to allow sports practices involving animals.

[4]Chapter 5.

[5]The Brazilian Constitution, in fact, has had more than 120 amendments, due, above all, to its length—it has 250 articles!

Environmental law also permeates other parts of the Constitution, such as the provision for *people's legal suit* (*ação popular*, that is a kind of *citizen suit* that only individual citizens can file) in defense of the environment (Art. 5, LXXIII); the distribution of environmental authority among the three federative entities (Art. 21 to Art. 30); the role of Public Prosecution Service (*Ministério Público*) in protecting the environment (Art. 129, III); the connection of the Economic Order to environmental protection (Art. 170, VI); the need for the exercise of the property right to take into account environmental preservation (Art. 186); the constitutional rules for the promotion and protection of Brazilian cultural heritage (Art. 215 to Art. 216-A); and the need to protect the environmental resources necessary for the well-being of the Indigenous populations (Art. 231, § 1).

Over and above, it was recognized that the right to an ecologically balanced environment is a fundamental right, protected from any future constitutional amendment. In other words, it is an *immutable* clause that cannot be removed from the original 1988 Constitution. Indeed, the Constitution established that some major points are not allowed to be revoked, altered, or eliminated, not even by constitutional amendment, which are considered to be *immutable clauses* (*cláusulas pétreas*) of the Constitution. Such clauses are intended to safeguard certain essential contents and to protect the Constitution against political manipulation and the absolutism of parliamentary majorities.[6] One of these clauses pertains to individual rights (Constitution, Art. 60, § 4, IV), which has been systematically interpreted to encompass all fundamental rights, including social, political, and the right to a healthy environment.[7]

All public authorities are bound by this fundamental right, and their actions, or lack thereof, must not cause disproportionate and illegitimate harm to the quality of the environment. In other words, "duties of protection" and "duties of implementation" are imposed on the government, which can even be deemed liable in case of unjustified failure to act in an adequate and timely manner.[8] Thus, one can indeed refer to an environmental or ecological Constitution in Brazil, which incorporates an "Ecological Rule of Law," although the traditional view of environmental law remains excessively anthropocentric, allowing for the overexploitation of nature and the destruction of the natural foundations of life.[9]

[6] Sarlet (2015), p. 431.

[7] On the scope of the *stony clause* in the sphere of fundamental rights, read: Sarlet (2015), pp. 440–450. Regarding the inclusion of the fundamental right to the environment as an *immutable clause* of the Constitution, consider the following: Fensterseifer and Sarlet (2019), pp. 112–114.

[8] On the binding of public authorities to the constitutional duties of environmental protection, see: Fensterseifer and Sarlet (2019), pp. 94–96.

[9] On the greening of the Brazilian Constitution and the formation of an ecological constitutional law, consider reading: Fensterseifer and Sarlet (2019). For a critical view and the need for a "revolution in current law" to provide true "ecological justice," see: Morato Leite and Silveira (2020), pp. 89–139.

4.3 Environmental Protection and Indigenous Peoples

A topic of great relevance to Brazilian environmental law is its connection with Indigenous peoples and their lands. According to a recent study, Indigenous and traditional peoples are responsible for safeguarding approximately one-third of Brazil's forests. Over the past 35 years, Indigenous Territories have preserved 20% of the nation's total forested area.[10]

The 1988 Constitution dedicates a specific chapter to Indigenous rights, addressed in Arts. 231 and 232. Article 231 recognizes Indigenous peoples' social organization, customs, languages, beliefs, and traditions, as well as their original rights to the lands they have traditionally occupied. It is the responsibility of the Federal Government to demarcate these lands, protect them, and ensure respect for all their assets.

Pursuant to § 1 of Art. 231, lands traditionally occupied by Indigenous peoples are those they permanently inhabit, use for their productive activities, require for the preservation of environmental resources essential to their well-being, and need for their physical and cultural reproduction, in accordance with their customs, traditions, and ways of life. Furthermore, lands traditionally occupied by Indigenous peoples are designated for their permanent possession, granting them exclusive usufruct rights over the soil's resources, as well as the rivers and lakes within these lands (§ 2).

The utilization of water resources, including hydroelectric potential, as well as the exploration and extraction of mineral resources on Indigenous lands, may only be carried out with the authorization of the National Congress and after consulting the affected communities. These communities are guaranteed participation in the results of the extraction, as provided by law (§ 3).

Regarding Indigenous lands, the Constitution establishes that they are inalienable and unavailable, and the rights over them are indefeasible. Moreover, the removal of Indigenous groups from their lands is prohibited, except through *ad referendum* of the National Congress, in cases of catastrophe or epidemic posing a threat to their population, or in the interest of national sovereignty, following deliberation by the National Congress. In any such scenario, the immediate return of the groups is guaranteed as soon as the risk ceases. Additionally, § 6 of Art. 231 stipulates that acts involving the occupation, ownership, or possession of the Indigenous lands, or the exploitation of natural resources in the soil, rivers, or lakes within these lands, are null and void and produce no legal effects. Exceptions are made for cases of significant public interest to the Federal Government, as determined by Supplementary Law. The nullity and extinction of such acts do not entitle compensation or legal actions against the Government, except as provided by law, in relation to improvements resulting from good-faith occupation.

It is worth mentioning that, according to the Constitution, Indigenous peoples, their communities, and organizations have standing to bring actions in court to

[10] Instituto Socioambiental (2022).

defend their rights and interests, with the Public Prosecutor's Office intervening in all stages of the proceedings (Art. 232).

Regarding the demarcation of Indigenous lands, there is an ongoing debate about the existence of a *temporal framework* (*marco temporal*) for the occupation of these lands as a prerequisite for recognizing a specific area as being of original Indigenous occupation. Land is the central element of Indigenous issues, particularly the profound and intimate relationship Indigenous peoples have with it.

The recognition of Indigenous lands has existed since the 1934 Constitution. In the 1988 Constitution, in addition to establishing that lands traditionally occupied by Indigenous peoples are federal property (Art. 20, XI), Arts. 231 and 232 introduced specific protection rules. Since the 1988 Constitution, two conflicting theories have emerged regarding the demarcation of Indigenous lands: the "*Temporal Framework*" (*Marco Temporal*) or "*Indigenous Fact*" (*Fato Indígena*) theory and the "*Indigenate*" (*Indigenato*) theory.

According to the first theory, Indigenous peoples would only have the right to the recognition of their original possession of lands if they were already occupying those lands at the time of the Constitution's promulgation on October 5, 1988. However, under the *Indigenate* theory, it is possible to recognize the Indigenous peoples' immemorial possession of a particular area, regardless of the date they were forced to leave those lands.

The matter was first decided by the Supreme Court in 2009 in a ruling that recognized the existence of the *Temporal Framework*. In summary, it was held that the Constitution established a specific date—its own promulgation date (October 5, 1988)—as an irreplaceable reference point for determining the occupation of a given geographic area by a particular Indigenous group. In other words, this date serves as the benchmark for recognizing the original rights of Indigenous peoples over the lands they have traditionally occupied. The Court further clarified, however, that the traditional nature of Indigenous possession is not forfeited in cases where, at the time of the Constitution's promulgation, reoccupation had not occurred solely due to the *persistent dispossession* (*renitente esbulho*) of those lands by non-Indigenous individuals. *Persistent dispossession* would be characterized by the existence of a possessory conflict, whether factual or litigated, at the time of the promulgation of the 1988 Constitution.[11] However, the matter had not yet been settled, as the same Court decided that the ruling issued at the time would have binding effects on other cases.

The debate was only revisited in 2023, when the Supreme Court, with a different composition, ruled differently. It held that the constitutional protection of Indigenous peoples' original rights over the lands they traditionally occupy is not contingent upon the existence of a temporal framework on October 5, 1988, nor on the occurrence of persistent dispossession, whether as a physical conflict or an ongoing legal dispute at the time of the Constitution's promulgation. This time, the Court's

[11] Brazilian Supreme Federal Court, Pet. 3388, Rapporteur Min. Carlos Ayres Britto, j. 19 Mar. 2009.

decision had binding effects on the entire Judiciary and Executive branches, requiring that future cases align with this new understanding.[12]

In this latest ruling, the Court acknowledged the significant difficulty that would be imposed on Indigenous peoples, given their vulnerability, if the *Temporal Framework* theory were upheld. The Court also affirmed, among other points, that the traditional occupation of Indigenous lands is compatible with the constitutional protection of the environment, ensuring the continued practice of Indigenous peoples' traditional activities.

The Court also addressed issues concerning the legal certainty of long-standing good-faith possessors of these lands. It ruled, in summary, that in the absence of traditional Indigenous occupation at the time of the Federal Constitution's promulgation, legal acts and transactions based on just title or good-faith possession of lands traditionally occupied by Indigenous peoples are valid and effective. Such individuals are entitled to fair and prior compensation from the Federal Government for necessary and useful improvements. Additionally, when the resettlement of these individuals is unfeasible, they shall be entitled to compensation from the Federal Government equivalent to the value of the undeveloped land.

Shortly after the Supreme Court's decision, however, the National Congress, in a clear instance of legislative backlash, passed Law n. 14,071/23 to establish precisely the opposite of what the Court had ruled, effectively reinstating the *Temporal Framework* approach. The approval of the new statute was clearly the result of pressure from agricultural sectors and opposition parties to the current government.

The controversy remains unresolved, as new actions for unconstitutionality have been filed against Law n. 14,071/2023, which are still pending judgment by the Supreme Court. In April 2024, Justice Gilmar Mendes ordered the nationwide suspension of judicial proceedings concerning the constitutionality of this law until the Supreme Court delivers a definitive ruling on the matter. Furthermore, the Justice initiated a conciliation process, granting the parties involved time to submit proposals for resolving the political and legal impasse related to the recognition, demarcation, use, and management of Indigenous lands.

In the current scenario, the legal situation of Indigenous peoples remains precarious. Indigenous reserves, with their undeniable importance in environmental protection, continue to be vulnerable to measures that may diminish their safeguards.

4.4 Legal Regime of Environmental Assets in Brazilian Law

In Brazil, environmental assets cannot be considered private property, nor the property of the state. This statement is taken directly from Art. 225 of the Constitution, which states that an ecologically balanced environment is to be regarded as a

[12]Brazilian Supreme Federal Court, RE 1.017.365/SC, Rapporteur Min. Edson Fachin, j. 27 Sept. 2023.

good of common use to all people and essential to a healthy quality of life. Before the 1988 Constitution, e.g., Brazilian Law provided for state ownership of wild fauna[13] and the existence of private forests.[14] Gradually, environmental assets have become considered "assets of common use of the people,"[15] that is, assets that do not belong to any private property, nor to the government, but assets for the use of the entire community, or assets necessary and important for the existence of all living beings, which cannot or should not be subject to the private use of anyone.[16]

The category of "assets for the common use of the people" was already mentioned in the earlier Civil Code of 1916 and was reaffirmed in the current Civil Code of 2002, which includes the following under this classification: rivers, seas, roads, streets, and parks. The 1965 Forest Protection Act (no longer in force today), as well as the current 2012 Act, have stated that forests within the national territory and other forms of vegetation are "assets of common interest to all the inhabitants of the country." Finally, the consideration of environmental goods as "goods for the common use of the people" has been definitively enshrined in Art. 225 of the 1988 Constitution.

Therefore, the environmental balance is featured as a *public good*, in the economic sense of the term, and environmental assets are *common resources* (*commons*),[17] whose protection raises *collective action problems*,[18] that usually are not sufficiently solved by the free interaction of economic agents within a market economy.[19] The definition of environmental goods and ecosystem services as *public goods*, which are for the benefit of everyone regardless of their individual contribution to the promotion or preservation of these goods, unfortunately, opens doors for opportunistic behavior or *free-rider issues*. Likewise, this reality favors the

[13] The Fauna Protection Act stipulated that wild animals are the property of the state (Law n. 5.197/67).

[14] The first Forest Protection Act of 1934 (Dec. 23.793/34), currently revoked, referred to "private forests."

[15] In Portuguese: "*bem de uso comum do povo.*"

[16] Justen Filho (2014), p. 1122.

[17] "*Public Goods,*" in economic sense, are those goods whose consumption cannot be excluded (the supplier cannot prevent consumption of the good by people who do not pay for it) or rivaled (more than one person can consume the same unit of the good at the same time, without the use by one preventing the use by another). "*Common Resources*" (or "*commons*") are those goods whose consumption cannot be excluded, but is rivalized, that is, their use by one person reduces or impedes the use by another, which is why it is a limited or scarce resource (Krugman and Wells 2009, pp. 459–478).

[18] A collective action problem is a scenario in which there is conflict between the individual interest and the group interest. In the scenario, each individual in the group faces a choice to either act selfishly or cooperate.

[19] In Bugge's words, one of the fundamental challenges in environmental law is that "many environmental goods and services are public goods in the economic sense." "Since they are not subject to property rights and do not have a price in the market, they also have at the outset a weaker legal status than various types of property rights and easily identified economic interests. Without public law protection, they may easily be harmed, reduced or destroyed" (Bugge 2013, pp. 13–12).

extraction of wealth through the exploitation of natural resources (or through the imposition of negative outputs) with no benefit for society. In economics, it is called *rent-seeking*. Individuals find no incentive to pay for the exploitation or protection of these assets and services, favoring themselves freely at the expense of what was paid or provided by others. And because natural resources are also considered *common resources* (*commons*), they can reach their exhaustion threshold in the case of unregulated collective use, as the chronicle of the *tragedy of the commons*[20] has well demonstrated.

In this regard, Morato Leite asserts that environmental goods are not *public property*, nor state-owned, but rather *goods of public interest*.[21] This perspective deprives the owner of the right to deplete or exploit natural resources unsustainably or to degrade ecological systems, due to the constitutional provision that considers the environment as a *macro-good* whose ownership belongs to everyone indiscriminately.[22]

Environmental goods, therefore, gain autonomy in relation to public and private property, and their definition as *public goods* and *common resources* demands adequate state regulation to avoid overexploitation and depletion. Consequently, rules and principles on the environmental function of property have been developed, as analyzed above,[23] emphasizing that its use must adhere to limits related to environmental preservation. An outcome of this recognition is the possibility of claiming compensation for pure ecological damages.

References

Bugge HC (2013) Twelve challenges in environmental law. In: Voigt C (ed) Rule of law for nature: new dimensions and ideas in environmental law. Cambridge University Press, New York, pp 3–26

da Silva JA (2010) Direito urbanístico brasileiro, 6th edn. Malheiros, São Paulo

Fensterseifer T, Sarlet IW (2019) Direito Constitucional Ecológico, 6th edn. Revista dos Tribunais, São Paulo

Hardin G (1968) The tragedy of the commons. Science 162(3859):1243–1248

Instituto Socioambiental (2022) Estudo comprova que Povos Indígenas e Tradicionais são essenciais para a preservação das florestas. https://www.socioambiental.org/noticias-socioambientais/estudo-comprova-que-povos-indigenas-e-tradicionais-sao-essenciais-para#:~:text=A%20maior%20concentra%C3%A7%C3%A3o%20se%20d%C3%A1,apenas%20duas%20Terras%20Ind%C3%ADgenas%20demarcadas. Accessed 18 Nov 2024

Justen Filho M (2014) Curso de Direito Administrativo, 10th edn. Revista dos Tribunais, São Paulo

Krugman P, Wells R (2009) Economics, 2nd edn. Worth Publishers, New York

Morato Leite JR (2015) Sociedade de Risco e Estado. In: Canotilho JJG, Morato Leite JR (eds) Direito Constitucional Ambiental Brasileiro, 6th edn. Saraiva, São Paulo. Kindle Edition

[20] Hardin (1968).

[21] Morato Leite (2015), position 3052–3060.

[22] Morato Leite (2015). Position 3031.

[23] Section 3.3.

Morato Leite JR, Silveira PG (2020) A ecologização do Estado de Direito: uma ruptura ao direito ambiental e ao antropocentrismo vigentes. In: Morato Leite JR (coord) A ecologização do direito ambiental vigente, 2nd edn. Lumen Juris, Rio de Janeiro, pp 89–139

Sarlet IW (2015) A Eficácia dos Direitos Fundamentais: uma teoria geral dos direitos fundamentais na perspectiva constitucional, 12th edn. Livraria do Advogado, Porto Alegre

Thomé R (2017) Manual de Direito Ambiental, 7th edn. JusPodivm, Salvador

United Nations Environment Programme (2019) Environmental Rule of Law: First Global Report. Nairobi, 2019. http://wedocs.unep.org/xmlui/bitstream/handle/20.500.11822/27279/Environmental_rule_of_law.pdf?sequence=1&isAllowed=y. Accessed 18 Nov 2024

Chapter 5
Animal Protection and Animal Rights

5.1 Introduction

The Brazilian legal system has also adopted protective rules for animals. In fact, an intrinsic value and dignity inherent to non-human life have been recognized, and many scholars even acknowledge the possibility of recognizing animals as subjects of rights.[1]

International law has, through several acts, vested intrinsic value to biodiversity and ecological systems regardless of their direct utility to humans, e.g., the 1992 Convention on Biological Diversity, the 1991 Protocol on Environmental Protection amending the 1959 Antarctic Treaty, and the 1978 Universal Declaration of Animal Rights.

It is also worth mentioning the influence of what has been called "new Latin American constitutionalism" upon the progressive greening of law. Primarily experienced in South American legal systems, this model arises from social demands from groups historically excluded from the decision-making process, notably Indigenous people,[2] and aims to restore the relationship between popular sovereignty and government that has been lost or never fully achieved.[3] This new constitutionalism is pervaded by the concept called *buen vivir* (good living), which, in the words of Marco and Wenczenovicz, "highlights the strengthening of community and solidarity relations, the common spaces and the most diverse ways of living collectively, respecting diversity and nature."[4]

---

[1] The discussion on the granting of rights to animals and ecosystems, in Brazilian and comparative law, is still extensive and will not be deepened in this book. For further reading on the subject, see: Benjamin (2011); Fensterseifer and Sarlet (2019), pp. 82–83, 138–191. For a defense of animals as subjects of fundamental rights in Brazilian constitutional law, see: de Paula Ataíde Jr (2020).

[2] Alves (2012).

[3] Dalmau and Pastor (2010).

[4] de Marco and Wenczenovicz (2018), p. 6.

R. Martins Costa Moreira, G. Wedy, *Brazilian Environmental and Climate Change Law*, https://doi.org/10.1007/978-3-031-87998-2_5

The new Latin American constitutionalism and the idea of *buen vivir* are infused with the concept of legal ecocentrism, especially the Ecuadorian Constitution of 2008, the first to grant in its text rights to nature (or *Pachamama*), and the Bolivian legislation, especially with the *Ley de Derechos de la Madre Tierra* of 2010.[5] Also in South America, it is worth mentioning the "Universal Declaration of the Rights of Mother Earth" of 2010, adopted at the "World People's Conference on Climate Change and the Rights of Mother Earth," held in Cochabamba, Bolivia. Along the same lines, decisions of courts in Colombia,[6] Argentina, India, and other countries have embraced the thesis that animals and nature are entitled to their own rights and can be defended in court.[7]

5.2 Animal Protection in Brazilian Law

In Brazil, the legal protection of fauna has a long history, and the Fauna Protection Act of 1967[8] provided, albeit in an incipient way, legal instruments on the subject. It was with the 1988 Constitution, however, that the protection and recognition of the animals' inherent dignity, independent of any benefit to humans, gained prominence in Brazilian law. The Constitution, thus, has required the government: (a) to preserve and restore the essential ecological processes and provide for the ecological treatment of species and ecosystems; (b) to preserve the diversity and integrity of the genetic patrimony of the country and to control entities engaged in research and manipulation of genetic material; (c) and, mainly, to protect the fauna and the flora, with prohibition, in the manner prescribed by law, of all practices which represent a risk to their ecological function, cause the extinction of species or subject animals to

---

[5]For an analysis of the political substratum that led to the constitutional and legislative provision of rights to nature in Ecuador and Bolivia, especially its relation to the "right to a good life" (*buen vivir*), see: Fatheur (2011). On the innovations brought by the Ecuadorian and Bolivian constitutions, linked to the idea of the *buen vivir*, in diversity and harmony with nature, see: Morato Leite and Silveira (2020), pp. 89–139.

[6]The Colombian Constitutional Court, in 2016, attributed to *Rio Atrato* (Atrato River) the status of the subject of right (*Sentencia T-622/16*. https://www.corteconstitucional.gov.co/relatoria/2016/t-622-16.htm). In 2018, the Supreme Court of Colombia did the same for the Colombian Amazon rainforest in a lawsuit against deforestation (Colombia 2018).

[7]For an overview of comparative law experiences in granting rights to nature or recognizing the environment as a subject of rights, see: Fensterseifer and Sarlet (2019), pp. 82–83; 129–198. For an analysis of the *eco-legal* order and ecological jurisprudence in foreign law, with examples from Argentina, India, Ecuador, Colombia, Belize, Philippines, New Zealand, Australia, and Benin, among others, see: Morato Leite and Silveira (2020), pp. 89–139.

[8]Law n. 5.197/67.

cruelty.[9] We should also mention the Environmental Crimes Act of 1998,[10] which has an entire chapter on offenses defined as "crimes against fauna."[11]

This constitutional and legislative framework provided the Brazilian Supreme Federal Court with grounds to award holdings considering unconstitutional sporting activities that result in cruelty to animals, such as *cockfights*,[12] *farra do boi*[13] (*ox spree*), and, more recently, *vaquejada* (another sort of ox spree).[14] However, regarding *vaquejada*, communities and groups that disagreed with the Supreme Court's decision and wished to continue this activity pressured the National Congress to allow its practice. This was due not only to the cultural roots of this activity, but also to the economic benefits it brings to the Northeastern Region.

In the face of an authentic *backlash effect* to this holding, Law n. 13.364/16 was approved, which recognizes certain activities with animals, among them the *Vaquejada*, as national cultural manifestations and brings these activities to the status of intangible assets that are part of the Brazilian cultural heritage and provides for the protection of animal welfare. Subsequently, the National Congress approved the Constitutional Amendment n. 96 of 2016, which added § 7 to Art. 255 of the Constitution. This provision states that sporting activities involving animals are not considered cruel, as long as they are cultural expressions, registered as intangible assets of Brazilian cultural heritage, and must be regulated by specific legal measures that ensure the welfare of the animals involved. This is not the end of the story yet. A new claim for judicial review was filed against this constitutional amendment, still pending judgment by the Supreme Court.[15]

These cases demonstrate the existence of sharp conflicts between cultural and economic interests on the one side, and the protection of animal wellbeing on the other. So far, the Brazilian Supreme Court has ruled that constitutional provisions for the preservation of the environment and the protection of animals against cruelty take precedence, imposing limits on cultural and sporting events. Whether the Supreme Court will maintain the same understanding in view of this constitutional amendment is still uncertain.

---

[9] Brazilian Constitution, Art. 225, § 1, I, II e VII.

[10] Law n. 9.605/98.

[11] Law n. 9.605/98, Art. 29–37.

[12] Brazilian Supreme Federal Court, ADI 1856, Rapporteur Min. Celso de Mello, j. 26 Mai 2011; Brazilian Supreme Federal Court, ADI 3776/RN, Rapporteur Min. Cezar Peluso, j. 14 Jun. 2007; Brazilian Supreme Federal Court, ADI 2514/SC, Rapporteur Min. Eros Grau, j. 29 Jun. 2005.

[13] *Farra do boi* (ox spree) is a typical ritual from the coast of the Brazilian state of Santa Catarina, which consists in releasing an ox on the street and then beating it or "making fun" of it by making it run after the people who were participating in the practice. It was considered unconstitutional by Brazilian Supreme Federal Court: RE 153531, Rapporteur Min. Marco Aurélio, j. 03 Jun. 1997.

[14] *Vaquejada* is a cultural activity from the Brazilian Northeast, considered to be a sport, in which two cowboys on horseback must pull down an ox by its tail. The practice was ruled unconstitutional by the Brazilian Supreme Federal Court (ADI 4983, Rapporteur Min. Marco Aurélio, j. 06 Oct. 2016).

[15] Brazilian Supreme Federal Court, ADIn 5728, pending.

Other relevant cases decided by the Supreme Court should be emphasized (to mention a few): (a) the Court has upheld the constitutionality of a state statute that forbids the use of animals for the development, experiments, and testing of cosmetic, personal hygiene, perfume, and cleaning products. Aside from the absence of a federal statute on the subject, this legislation establishes a level of fauna protection that is higher than that of the Federal Government, within the supplementary constitutional jurisdiction of the state;[16] (b) however, the Court has found constitutionally valid a state law that, to protect religious freedom, allows the ritual sacrifice of animals in cults of religions of African origin;[17] (c) the Supreme Court has decided that it is unconstitutional the interpretation of the federal legislation that allows the immediate slaughter of seized animals in mistreatment situations.[18]

The Superior Court of Justice similarly accepted the idea that animals have intrinsic value and that their protection can be ensured independently of human rights. In a case it decided, the Court ruled that a wild bird should remain with the person it had been with for decades, despite the unlawful nature of that possession.[19]

Therefore, it is possible to say that Brazilian law welcomes an *ecocentric perspective*, still in development, by admitting, at least, that animals and non-human life deserve legal protection, regardless of any benefit to human rights. In this perspective, animals have their own dignity and intrinsic value, which may pose limits to individual rights.

The notion that animals or nature are entitled to rights or legal standing, with the possibility of filing a lawsuit on their own behalf, as seen in other legal systems, is not yet established or widely accepted. Additionally, there is no clear guidance from the jurisprudence of the Superior Court of Justice or the Supreme Court on this matter. In any case, there are subnational laws recognizing fundamental rights for animals or non-human life, such as the *Code of Animal Rights and Welfare of the State of Paraíba* and the provisions for rights of nature in municipal legislation of Cáceres,[20] and as well as a state Court decision to that effect.[21]

---

[16] The most recent ruling is: Brazilian Supreme Federal Court, ADI 5995, Rapporteur Min. Gilmar Mendes, j. 27 Mai 2021. In the same vein: Brazilian Supreme Federal Court, ADI 5996, Rapporteur Min. Alexandre de Moraes, j. 15 Apr. 2020.

[17] Brazilian Supreme Federal Court, RE 494601, Rapporteur Min. Edson Fachin, j. 28 Mar. 2019.

[18] Brazilian Supreme Federal Court, ADPF 640 MC-Ref/DF, Rapporteur Min. Gilmar Mendes, j. 17 Sept. 2021.

[19] Superior Court of Justice, REsp 1797175/SP, Rapporteur Min. Og Fernandes, j. 28 Mar. 19.

[20] Cáceres is a city of the state of Mato Grosso do Sul.

[21] The decision was issued by the Paraná State Court of Appeals, which recognized the legitimacy of animals to initiate a lawsuit on their own behalf (AI n. 0059204-56.2020.8.16.0000). To read a news article about the case: The RIO Times (2021).

References

Alves MV (2012) Neoconstitucionalismo e novo constitucionalismo latino-americano: características e distinções. Revista da Seção Judiciária do Rio de Janeiro 19–34:133–145

Benjamin AHV (2011) A natureza no direito brasileiro: coisa, sujeito ou nada disso. NOMOS: Revista do Programa de Pós-Graduação em Direito da UFC 31:79–96

Colombia (2018) Corte Suprema de Justicia. *Corte* Suprema ordena protección inmediata de la Amazonía Colombiana. http://www.cortesuprema.gov.co/corte/index.php/2018/04/05/corte-suprema-ordena-proteccion-inmediata-de-la-amazonia-colombiana/. Accessed 18 Nov 2024

Dalmau RM, Pastor RV (2010) Aspectos generales del nuevo constitucionalismo latino-americano. Corte Constitucional de Ecuador para el período de transición. El nuevo constitucionalismo en América Latina. Quito: Corte Constitucional del Ecuador, 2010, pp 9–43. https://files.cercomp.ufg.br/weby/up/14/o/34272355-Nuevo-Constitucionalismo-en-America-Latina.pdf?1352144011. Accessed 18 Nov 2024

de Marco CM, Wenczenovicz (2018) Lições de 'Buen Vivir': impactos ambientais sobre as comunidades indígenas no Brasil contemporâneo. Revista Brasileira de Direito 14(2):1–8

de Paula Ataíde V Jr (2020) Direito Animal e Constituição. Revista Brasileira de Direito e Justiça 4: 13–68

Fatheur T (2011) Buen vivir: a brief introduction to Latin America's new concepts for the good life and the rights of nature. Fundação Heinrich Böll. https://www.boell.de/sites/default/files/endf_buen_vivir_engl.pdf?dimension1=division_iup. Accessed 18 Nov 2024

Fensterseifer T, Sarlet IW (2019) Direito Constitucional Ecológico, 6th edn. Revista dos Tribunais, São Paulo

Morato Leite JR, Silveira PG (2020) A ecologização do Estado de Direito: uma ruptura ao direito ambiental e ao antropocentrismo vigentes. In: Morato Leite JR (coord) A ecologização do direito ambiental vigente, 2nd edn. Lumen Juris, Rio de Janeiro, pp 89–139

The RIO Times. Dogs conquer right to take legal action against former owners in Brazil. Sept. 16, 2021. https://www.riotimesonline.com/brazil-news/rio-politics/dogs-conquer-right-to-take-legal-action-against-former-owners-in-brazil/. Accessed 18 Nov 2024

Chapter 6
Environmental Federalism

6.1 Introduction

The evolution of environmental law in Brazil has sparked important debates regarding the distribution of legislative and administrative authority across the three tiers of the federation. In this issue, the Brazilian Constitution has vested with jurisdiction[1] all federative entities to ensure and achieve ecological protection. The power to legislate and administer environmental matters is entrusted to the three levels of the Federation: federal, state, and municipality. It is important to note that the Federal District (*Distrito Federal*), where the capital of the country is located, is a *sui generis* entity, which encompasses state and municipal competences.[2]

In general, our Constitution adopts the technique of enumerating the powers of the Federal Government (Arts. 21 and 22), with remaining competence for the states (Art. 25, § 1) and specific duties for the Municipalities (Art. 30). The Constitution also allows for the delegation of legislative authority from the National Congress to the states (Art. 22, sole paragraph), and establishes matters assigned jointly to all federative levels (Arts. 23 and 24).[3]

The 1988 Constitution distinguishes basically two types of competence that are assigned to the three levels of the Federation: (a) *material* or *administrative competence* is related to the implementation of environmental policy, by means of managing, organizing, maintaining, and executing its own affairs and duties, within the limits established by the Constitution.[4] Examples of material or administrative

[1] In Brazil, we refer *competência* (competence) to indicate the power of a certain authority or federative entity.

[2] According to Art. 32 of the Constitution.

[3] da Silva (2014), p. 477. In Silveira's perspective, the 1988 Constitution preserved the two methods of cutting jurisdiction: (a) the horizontal cut that embodies the private or exclusive powers; (b) the vertical cut that characterizes the common and concurrent powers (Silveira 2005, pp. 61–62).

[4] Bulos (2022), pp. 1002–1003.

R. Martins Costa Moreira, G. Wedy, *Brazilian Environmental and Climate Change Law*, https://doi.org/10.1007/978-3-031-87998-2_6

authority in environmental law can include the power of granting an environmental license, monitoring environmental violations and issuing administrative orders and penalties, and establishing conditions for economic activities and administrative limitations; (b) *legislative competence* is considered as the power of the political entity to issue imperative, general and abstract rules, based on the limits set by the Constitution.[5]

Within the domain of environmental law, all three levels of the Federation hold both administrative and legislative powers (Constitution, Art. 23, III, IV, VI, VII, and XI; Art. 24, VI, VII, and VIII), as we shall see in the coming chapters.

6.2 Material or Administrative Competence

Article 23 of the Brazilian Constitution included within the administrative competence of the three tiers of the Federation: to protect documents, works, and other assets of historical, artistic or cultural value, monuments, remarkable landscapes, and archeological sites (III); to prevent works of art and other assets of historical, artistic and cultural value from being taken out of the country, destroyed or from being deprived of their original characteristics (IV); to protect the environment and fight pollution in any of its forms (VI); to preserve the forests, fauna, and flora (VII); register, monitor and control the concessions of rights to research and exploit water and mineral resources within their territories (XI).

This means that the Federal, State, Federal District, and Municipal governments have equal authority to perform the acts listed above. The authority to exercise administrative powers, and consequently to oversee compliance with environmental legislation, is not confined to a specific public entity. This applies regardless of the location of the harm or threat, the ownership of the property, or the licensing authority involved. Along these lines, the Superior Court of Justice has adopted the concept that there is no question of exclusive power to one or another federative entity (national or subnational) to promote protective measures and enforce environmental law. The Constitution establishes a comprehensive supervisory framework to be implemented by all levels of the Federation, regardless of where the threat or damage is occurring, as well as the public entity vested in power to grant an environmental license.[6]

In addition to these powers assigned equally to the three federative spheres, the 1988 Constitution also specifies areas of exclusive authority of the Federal Government, remaining or residual competence of the states, and competence for matters of predominantly local interest to the municipalities.

[5] Bulos (2022), pp. 1002–1003.

[6] Superior Court of Justice, AgInt no AREsp 1458422/SP, Rapporteur Min. Assusete Magalhães, j. 17/12/2019; AgInt no AREsp 1499874/SC, Rapporteur Min. Mauro Campbell Marques, j. 19/11/2019; AgRg no REsp 1.417.023/PR, Rapporteur Min. Humberto Martins, e.g.

6.2.1 Exclusive Administrative Competence of the Federal Government

The 1988 Constitution, under Art. 21, includes in the realm of exclusive material (or administrative) competence of the Federal Government the following functions related to environmental law: prepare and carry out national and regional plans for land management and economic and social development (IX); operation of electric power services and facilities and the energetic exploitation of watercourses, jointly with the States wherein those hydroenergetic potentials are located (XII, b); establish a national system for the management of hydric resources and define criteria for the concession of the right to their use; (inc. XIX); establish directives for urban development, including housing, basic sanitation, and urban transportation; (XX); operate nuclear energy services and facilities of any nature and exercise state monopoly over research, mining, enrichment and reprocessing, industrialization and trade in nuclear ores and their by-products (XXIII).

The exclusive powers of the Federal Government mean that subnational entities cannot act directly on the subject, except when it also involves matters of common competence (CF, Art. 23), as shall be analyzed below.

6.2.2 Remaining or Residual State Authority

For the states, the Brazilian Constitution grants remaining or residual competence. It refers to powers that are not prohibited to them (Art. 25, § 1), consisting of functions that belong neither to the Federal nor to the local government (Arts. 21 and 30), and nor to all entities in common (Art. 23). Residual jurisdiction refers to the authority retained by the states to address or legislate on any matters not expressly assigned to another entity, whether federal or municipal.

Exceptionally, the Constitution enumerates for the states some functions, which, in what concerns environmental law, are: (a) to operate, directly or by means of concession, the local services of piped gas (Art. 25, § 2); and (b) to establish metropolitan regions, urban agglomerations and microregions, formed by grouping of adjacent municipalities, in order to integrate the organization, the planning and the operation of public functions of common interest (Art. 25, § 3). In any case, states also hold the duty (granted to all federative entities by Art. 23) to act in defense of the environment by means of administrative measures.

6.2.3 Administrative Competence of Municipalities

In addition to the shared competencies outlined in Art. 23, municipalities have also been granted exclusive powers, among them “to promote, wherever pertinent, adequate land management, by means of planning and controlling of use, apportionment and occupation of the urban land" (Art. 30, VIII) and "to promote the protection of the local historic and cultural heritage, with due regard for federal and state legislation and supervision" (Art. 30, IX).

The discussion regarding the municipalities' supplementary competence and the power to legislate on matters of predominantly local interest will be addressed in due course.

6.2.4 Federative Cooperation and the Supplementary Law n. 140/2011

To solve or reduce federative conflicts in environmental matters, especially regarding the power to analyze and grant environmental licenses, Supplementary Law[7] (*Lei Complementar*) n. 140/11 was enacted. This statute regulates Art. 23, § 1 of the Constitution and establishes rules for cooperation among the federative entities for the exercise of administrative power in environmental matters. Its main objectives have been to promote environmental management decentralization, optimize public entities efforts, and increase legal certainty, to make environmental protection more effective.[8]

Article 4 of Supplementary Law n. 140/2011 states that the federative entities may use, among others, the following instruments of institutional cooperation: I—public partnerships;[9] II—covenants, technical cooperation agreements, and other similar tools with public agencies and entities; III—National Tripartite Commission, State Tripartite Commissions, and the Federal District Bipartite Commission; IV—public and private funds and other economic measures; V—delegation of functions from one federative entity to another; VI—delegation of the execution of administrative actions from one federative entity to another.

Pursuant to Art. 6 of Supplementary Law n. 140/2011, cooperative efforts among the federal government, states, the Federal District, and municipal governments must be cultivated to ensure sustainable development while harmonizing and integrating all governmental policies. This statute also establishes rules on the competence for administrative actions, notably for environmental licensing (Arts. 7, 8, 9, and 10), which shall be subject to further considerations.

---

[7] *Supplementary Laws* (*Leis Complementares*) differ from other *Ordinary Laws* (*Leis Ordinárias*) in that they require the vote of a majority of the members of the House of Representatives and the Federal Senate to be approved. They must be adopted to regulate specific subjects, when expressly determined in the Constitution of the Republic.

[8] Farias (2020), pp. 3–6.

[9] *Consórcios Públicos*, as the legal act states, is a kind of administrative tool of federative partnership to achieve specific public goals.

6.3 Legislative Competence

Legislative or formal competence pertains to the authority of federative entities to enact statutes. This power may be exclusive to a specific federative entity or concurrent among all of them.

6.3.1 *Exclusive Legislative Competence of the National Congress (Federal Government)*

Although the power to enact statutes for the environmental realm is, as a rule, concurrent to all federative levels, the Constitution includes cases of exclusive competence. Thus, in association with the exclusive material authority of the Federal Government mentioned above, exclusive legislative power is provided for Art. 22 of the Constitution, to deal with: water and energy (IV); mineral resources in general, and metallurgy (XII); Indigenous populations (XIV); and nuclear activities of any nature (XXVI).

The Supreme Court has been called upon to resolve several federative conflicts related to environmental legislation, especially when states and municipalities advance in areas of exclusive competence of the National Congress. Along these lines, e.g., the Supreme Court has decided that state legislation that regulates activities related to the nuclear sector at the regional level is unconstitutional, due to violation of the federal power to legislate on nuclear activities, which includes the authority to inspect the execution of these activities and to legislate on such inspection.[10] On the subject, more recently, the Supreme Court has declared that a state Constitution statute that provides for the disposal of atomic waste and the deployment of nuclear power plants is contrary to the Constitution.[11] This is because, as mentioned previously, only the Federal Government, through the National Congress, has the authority to regulate matters related to nuclear energy.

Supreme Court has also held for the unconstitutionality of state laws that, besides advancing on other issues that are within the domain of the National Congress (such as legislation on civil and commercial law, foreign and interstate commerce, the legal regime of ports, lake, river, sea, air and aerospace navigation, traffic and transport, and mineral resources, in the form of Art. 22 of the Constitution), also violate the rules of concurrent legislative competence of Art. 24. Based on these grounds, the Supreme Court has found that the following cases pose a violation of the constitutional rules on environmental federalism: (a) state legislation restricting the use and trade of imported pesticides;[12] (b) state rules establishing prohibitions on

---

[10] Brazilian Supreme Federal Court, ADI 1575, Rapporteur Min. Joaquim Barbosa, j. 07 Apr. 2010.

[11] Brazilian Supreme Federal Court, ADI 6909/PI, Rapporteur Min. Alexandre de Moraes, j. 17 Sept. 2021; ADI 6913/DF, Rapporteur Min. Alexandre de Moraes, j. 17 Sept. 2021.

[12] Brazilian Supreme Federal Court, ADI 3852, Rapporteur Min. Dias Toffoli, j. 07 Oct. 2015.

the cultivation, handling, importation, industrialization, and commercialization of genetically modified organisms (GMOs) in its territory.[13] On the other hand, it is equally unconstitutional for a state legislation to fully leave the regulation of cultivation and activities with GMOs to federal statutory law, since the state's ability to legislate on environmental matters cannot be waived;[14] c) state law that prevents the exercise of mining activities by individuals;[15] (c) state law governing mining activities, including the definition of relevant concepts, the designation of areas where such activities may occur, and the authorization of mercury usage under certain conditions.[16]

6.3.2 *Remaining or Residual Legislative Competence of the States*

The legislative competence of the states, in equal measure to what was mentioned above, is of a residual or remaining[17] nature, according to Art. 25, § 1 of the Constitution, when it prescribes that "all powers that this Constitution does not prohibit the states from exercising shall be conferred upon them."

Therefore, by assigning to the state powers that are not constitutionally forbidden, the 1988 Constitution embraced, simultaneously, the state competence of executive and legislative nature. Therefore, it is within the state's authority to legislate on any matter, including environmental issues, as long as these do not fall under the jurisdiction of the federal or municipal governments.[18]

6.3.3 *Concurrent Legislative Competence*

The power to legislate on environmental protection is, for the most part, and as a rule, concurrent between the three federative tiers.

---

[13] Brazilian Supreme Federal Court, Pleno, ADI 3035, Rapporteur Min. Gilmar Mendes, j. 06 Apr. 2005.

[14] Brazilian Supreme Federal Court, ADI 2303/RS, Rapporteur Min. Marco Aurélio, j. 05 Sept. 2018.

[15] Brazilian Supreme Federal Court, ADI 5077, Rapporteur Min. Alexandre de Moraes, j. 25 Oct. 2018.

[16] Brazilian Supreme Federal Court, ADI 6672/RR, Rapporteur Min. Alexandre de Moraes, j. 14 Sept. 2021.

[17] As we said above, remaining or residual power means the power held to remain at the disposal of the state authority after an enumeration or delegation of specified powers to other authorities (federal and municipal).

[18] Ferreira (2011), position 4373–4361.

Article 24 of the Brazilian Constitution establishes concurrent legislative powers for the federal government and the states, encompassing the following areas of legislation: "forests, hunting, fishing, fauna, nature preservation, defense of the soil and natural resources, environment protection and pollution control" (VI), "protection of the historic, cultural and artistic heritage, as well as of assets of touristic interest and landscapes of outstanding beauty" (VII); and "liability for damages to the environment, to consumers, to assets and rights of artistic, aesthetic, historical, and touristic value, as well as to remarkable landscapes" (VIII).

It is worth noting that, although Art. 24 of the Constitution lists only the Federal Government, the states, and the Federal District, a systematic interpretation of the Constitution, along with Art. 23, III, VI, and VII, Art. 30, II, and Art. 182, allows us to assign to the municipalities the legislative power for the issues cited therein.[19] Moreover, Art. 6, § 2 of Law n. 6938/81 establishes that the municipalities, observing federal and state rules and standards, can also enact statutes on environmental law.

In the context of the concurrent legislative competence, all federative entities have the power to legislate, but only the National Congress is responsible for enacting a law on a general provision, which applies to all national and subnational entities and conditions the validity of state and municipal laws (Constitution, Art. 24, §§ 1 to 4). Thus, concurrent competence encompasses two key elements: the ability of multiple federative entities to legislate on the same subject and the primacy of the National Congress in establishing general provisions (called *general laws* or *normas gerais*).[20]

The concept and limits of the *federal general statutory provision* in the field of concurrent legislative competence are ripe for controversy. A *general rule* is usually related to the "principles and general guidelines for a legislative action"[21] of other entities, a "framework standard,"[22] not exhaustive, which leaves room for the states and municipalities to legislate according to regional and local peculiarities.

However, it is important to note that the dynamism of the federative structure and the expanding role of the state in regulating economic and social matters leave little room for a definitive interpretation of the term *general law*, which contains an unclear legal concept or an *open-texture term* (*conceito jurídico indeterminado*), to be clarified in a case-by-case basis.[23]

On this point, the Supreme Court has struggled to draw guidelines to define the scope of action of *general federal provision* on the one hand, and state and municipal authority on the other. Thus, the Supreme Court, in several judgments, has stated that the key to federative success lies in balancing competition and cooperation. The

---

[19] Cappelli et al. (2013), p. 80; Branco and Mendes (2013), p. 806; Silveira (2005), p. 79.

[20] da Silva (2014), p. 481.

[21] da Silva (2010), pp. 64–65.

[22] Horta (1995), pp. 419–420; Brazilian Supreme Federal Court, ADI 927 MC, Rapporteur Min. Carlos Velloso, j. 03 Nov. 1993.

[23] de Freitas (2000), p. 322; Moreira (2016); Silveira (2005), p. 83.

entities of the federation must respect interdependence when exercising their functions.[24] The Court has also stated that, if, on the one hand, the general federal provision cannot prevent the exercise of state competence to supplement the matters listed in Art. 24, on the other hand, state legislation cannot be allowed to encroach on federal competence and legislate the matter in a manner contrary to general federal law, distorting the minimum of normative unity desired by the Constitution.[25]

The contrast between generality and specificity, thus, is linked to the nature of the subject to be legislated. It is important to highlight at this point that the impacts arising from degrading activities conducted within cities or states frequently extend beyond municipal, regional, and even national boundaries. Thus, in environmental law, it is reasonable for the National Congress to establish at least a threshold of protection in order to safeguard the welfare of the people and the protection of the right to an ecologically balanced environment.[26]

The delineation of concurrent legislative competence and the definition of *general provision* were the subjects of significant debate in the Supreme Court regarding the validity of a São Paulo State regulation that prohibits the use, sale, and production of asbestos-based products within its territory, despite the existence of a federal statute (Law n. 9055/95) that permits the use of this substance. According to the Court, the Federation needs to find a point of stability between centralization and decentralization. The concurrent power exercised by the Federal Congress should encompass the national interests, which cannot be limited to the state borders. Nevertheless, federal law cannot encompass every detail to ensure that substantial authority remains with the states. It was considered that, at present, the national legislation on the subject is no longer compatible with the Constitution, because allowing the use of asbestos violates the right to health and the environment. Consequently, the states now have full legislative power over the matter until a new federal law is enacted.

The Supreme Court also pointed out that the *process of unconstitutionalization* of Law n. 9.055/1995 was brought about by a change in the factual background, that is, there is currently a consensus of the highly carcinogenic nature of asbestos and the unfeasibility of its safe use. Asbestos can be replaced by other materials which do not have carcinogenic properties, and which are recommended by the federal agency ANVISA.[27] Consequently, the Supreme Court declared Art. 2 of Federal Law n. 9,055/1995 unconstitutional for violating the right to health (as established in Arts. 6 and 196 of the Constitution) and the state's obligation to mitigate work-related risks through health, hygiene and safety regulations (Art. 7, XXII), as well as

---

[24] Brazilian Supreme Federal Court, ADPF 24/DF, Rapporteur Min. Alexandre de Moraes, j. 22 Aug. 2019.

[25] Brazilian Supreme Federal Court, ADI 3357, Rapporteur Min. Dias Toffoli, j. 30 Nov. 2017.

[26] On the need for *general federal rules* that establish a minimum level of protection to be observed by all entities of the Federation, read on: Moreira (2016). This is regardless of how detailed such general rules may get: Farias (2003).

[27] ANVISA is the short of *Agência Nacional de Vigilância Sanitária* or *National Health Surveillance Agency*.

for posing risks to the environment (Art. 225). Due to the invalidity of the *general federal law* (Law n. 9.055/95), the state and municipal governments have full legislative power over the matter until new federal legislation is enacted. Since the state law in question prohibits the use of asbestos, it has been held valid by the Court.[28]

In another judgment on the same subject, the Supreme Court has decided that, in light of the scientific knowledge on the extent of the harmful effects of asbestos on health and the environment, and the evidence of the ineffectiveness of the control measures contemplated therein, tolerance for the use of chrysotile asbestos, as set forth in Art. 2 of Law n. 9.055/1995, does not adequately and sufficiently protect the fundamental rights to health and environment, nor does it align with the international obligations assumed by Brazil which have shaped the content of these rights, especially International Labor Organization Conventions n. 139 and 162 and the Basel Convention. The Supreme Court has ruled that insufficient government protection is unconstitutional, and that state and municipal laws concerning its regulation are valid, even if they result in the banning of all use of asbestos.[29]

Conflicts involving concurrent legislative competence are quite prevalent, with the Supreme Court often adjudicating these issues. To provide insight into the different types of conflicts that arise in this context and the interpretations given to the rules governing federative jurisdiction, we will briefly summarize some of the cases ruled upon by the Supreme Court below:

(a) The Court ruled the unconstitutionality of state and municipal laws conflict with the provisions of the Forest Code concerning permanent preservation areas;[30]
(b) In the context of environmental licensing and environmental impact assessment, the Supreme Court has also decided that the laws on this subject must be in accordance with constitutional rules and federal legislation. Because of this, the Court has held that a state rule that waives the obligation of carrying out environmental licensing and environmental impact assessment for potentially polluting activities is unconstitutional.[31] Moreover, the same Court has also found that, although states are entitled to regulate their regional affairs, they are not allowed to establish a type of *sole license*,[32] comprising all phases into one. On the other hand, the Court has decided that it is feasible for the states to

[28] Brazilian Supreme Federal Court, ADI 3937/SP, Rapporteur Min. Dias Toffoli, j. 24 Aug. 2017; ADI 3357, Rapporteur Min. Dias Toffoli, j. 30 Nov. 2017.

[29] Brazilian Supreme Federal Court, ADI 3470, Rapporteur Min. Rosa Weber, j. 29 Nov. 2017.

[30] Brazilian Supreme Federal Court, Pleno, ADI 4988, Rapporteur Min. Alexandre de Moraes, j. 19 Sept. 2018. More recently: ADI 5675/MG, j. 17 Dec. 2021.

[31] Brazilian Supreme Federal Court, ADI 6288, Rapporteur Min. Rosa Weber, j. 23 Nov. 2020; ADI 1086 MC, Rapporteur Min. Ilmar Galvão, j. 01 Aug. 1994.

[32] As we will have the opportunity to further discuss below, environmental licensing in Brazil is composed of three phases: preliminary license, installation license, and operation license. The Supreme Court held that it violates the rules of concurrent legislative competence when a state law provides for the hypothesis of a *sole license*, that covers all three phases in one: Brazilian Supreme Federal Court, ADI 5475, Rapporteur Min. Carmen Lúcia, j. 20 Apr. 2020.

provide simplified licenses for activities with low environmental impact, according to regional features;[33]

(c) Regarding the power to establish "environmental zoning" (land use regulation), the Supreme Court has decided that, although the state competence is not limited to the guidelines set forth in federal law, they must observe it, and, because of this, a state law on land use regulation that waives environmental impact assessment for activities with significant environmental impact is unconstitutional;[34]
(d) As mentioned above, the Supreme Court has held that a state rule prohibiting the use of animals for the development, experiments, and testing of cosmetic products, personal hygiene products, and perfume is valid. In this case, there is no breach of the National Congress's competence regarding general provisions, since the issuance of state rules which are more protective of the environment is permitted, based on their regional peculiarities and the preponderance of their interests;
(e) The Supreme Court has also held that the constitutional provision of the State of São Paulo that prohibits hunting in its respective territory does not violate federal legislation, except for controlled hunting and scientific hunting, because these types of hunting are in favor of balancing the ecosystem.[35]

Likewise, the municipal competence to supplement federal and state legislation is often questioned and can be a source of conflicts, a topic to be addressed in the next chapter.

6.3.4 Supplementary Legislative Competence of Municipalities

The municipalities' legislative competence is guided by the *predominance* of local interests. *Predominance* should not be conflated with *exclusivity*, as no public or environmental issue is solely local, regional, or national. Instead, there has always been an overlapping area among these levels of interest.[36] Thus, in numerous instances, determining the predominance of local interest—which grants municipalities legislative power—presents significant challenges and necessitates careful interpretation by the courts.

At this point, urban land use planning is predominantly a matter of local interest, as clearly stated in Art. 30, VIII. Art. 30, VIII and Art. 182 of the Constitution.

[33] Brazilian Supreme Federal Court, ADI 4615, j. 20 Sept. 2019; RE 1264738 AgR, j. 24 Aug. 20.

[34] Brazilian Supreme Federal Court, ADI 4069, Rapporteur Min. Edson Fachin, j. 08 Sept. 2020.

[35] Brazilian Supreme Federal Court, ADI 350/SP, Rapporteur Min. Dias Toffoli, j. 18 Jun. 2021.

[36] On the difference between predominance and exclusivity of local interest, and the difficulty in identifying the municipal interest zone, see: Meirelles (2008), p. 111; da Silva (2014), p. 478.

However, local legislation must adapt to federal legislation on the subject. This is the case of the City Statute (Federal Law n. 10.257/2001), enacted based on Art. 24, I of the Brazilian Constitution, to regulate Art. 182 and Art. 183, and the Urban Land Parcelling Act (Federal Law n. 6.766/79), which establishes minimum requirements for the apportionment of urban land, so that municipalities cannot violate its provisions.

Thus, the same applies here as was said about the concurrent legislative competence: municipal laws must not breach the limits of federal legislation (Brazilian Constitution, Art. 24, § 1), and in case it happens, the municipal law will have its effectiveness halted (Art. 24, § 4).

Regarding the protection of the environment, the Supreme Court has ruled that the municipality has concurrent power to legislate on the environment in joint authority with the federal and state governments, within the limits of its local interest, and provided that such laws are harmonious with the legislative framework of other federative entities. In this case, the Supreme Court held that the municipality lacks legislative power to prevent the use of straw burning during sugarcane harvest. Between the prohibition imposed by municipal law and the progressive elimination regulated by state legislation, the Supreme Court decided that the latter should prevail, because the planned and gradual phase-out of sugarcane burning does not concern predominantly local interests, and it is more compatible with constitutional law. No interpretation should be allowed that does not support the municipal interest in ensuring that its population enjoys a balanced environment. But, in this case, it is a question of identifying the preponderance of these notably regional interests.[37]

The Supreme Court also found that the municipalities can legislate, at the local level, on measures for the protection and defense of the environment and health, when ruled in favor of the validity of a municipal law imposing fines against environmental pollution resulting from the emission of smoke by motor vehicles in the urban perimeter.[38] It should also be noted that the Supreme Court has considered constitutional a municipal law that restricts the use of herbicides in certain locations and at certain seasons, a situation that differs from a complete and general ban on the use of these products.[39] The same Court has also ensured local autonomy by quashing a state constitutional rule that prohibits municipalities from changing the original destination, purposes, and objectives of land allotments defined as green or institutional areas.[40]

Accordingly, the examination of these precedents reveals that granting authority to the three levels of government in support of environmental protection— while enhancing the supervisory and protective framework for the environment and

[37] Brazilian Supreme Federal Court, Pleno, RE 586.224, Rapporteur min. Luiz Fux, j. 5 Mar. 2015.

[38] Brazilian Supreme Federal Court, Pleno, RE 194704/MG, Rapporteur Min. Edson Fachin, j. 29 Jun. 2017.

[39] Brazilian Supreme Federal Court, ARE 118110 AgR, Rapporteur Min. Gilmar Mendes, j. 11 Oct. 2019.

[40] Brazilian Supreme Federal Court, ADI 6602/SP, Re. Rapporteur Cármen Lúcia, j. 11 Jun. 2021.

enabling regional and local entities to address issues pertinent to their unique economic, social, and ecological contexts—has also resulted in numerous disputes. These disputes necessitate ongoing efforts for resolution by the Superior Court of Justice and the Supreme Court.

References

Branco PGG, Mendes GF (2013) Curso de Direito Constitucional, 8th edn. Saraiva, São Paulo

Bulos UL (2022) Curso de Direito Constitucional, 15th edn. Saraiva, São Paulo

Cappelli S, Marchesan AMM, Steigleder AM (2013) Direito Ambiental, 7th edn. Verbo Jurídico, Porto Alegre

da Silva JA (2010) Direito urbanístico brasileiro, 6th edn. Malheiros, São Paulo

da Silva JA (2014) Curso de direito constitucional positivo, 37th edn. Malheiros, São Paulo

de Freitas VP (2000) Mata Ciliar. In: de Freitas VP (ed) Direito Ambiental em Evolução. n. 2. Juruá, Curitiba

Farias T (2003) A Edificação Urbana à Margem de Rios e de Outros Reservatórios de Água em Face do Código Florestal. Fórum de Direito Urbano e Ambiental – FDUA 10:959–967

Farias T (2020) *Competência Administrativa Ambiental*: fiscalização, sanções e licenciamento ambiental na Lei Complementar 140/2011. Lumen Juris, Rio de Janeiro

Ferreira HS (2011) Competências ambientais. In: Canotilho JJG, Leite JRM (eds) Direito Constitucional Ambiental Brasileiro, 6th edn. Saraiva, São Paulo. Kindle Edition

Horta RM (1995) Estudos de Direito Constitucional. Del Rey, Belo Horizonte

Meirelles HL (2008) Direito Municipal Brasileiro, 16th edn. Malheiros, São Paulo

Moreira RMC (2016) Áreas de preservação permanente e urbanização: muitos conflitos e uma controvérsia. Interesse Público 18(99):90–116

Silveira PA (2005) Competência ambiental, 1st edn. 4. Tir. Juruá, Curitiba

Chapter 7
National Environmental Policy and National Environmental System

7.1 Introduction

The National Environmental Policy (NEP) was established by Law n. 6.938/81, marking a significant milestone in Brazilian environmental law and the legal protection of nature. It serves as a crucial instrument for planning and stability within Brazilian environmental policy, offering a holistic and integrated perspective on environmental law while moving away from the improvisation and fragmentation of prior efforts.[1] Although, there is still much to evolve in terms of environmental protection in Brazil.

Law n. 6.938/81 has brought, thus, a systematic approach of the Brazilian environmental law, without, however, reaching the status of a complete and comprehensive *Environmental Law Code*. Article 2 states that the National Environmental Policy aims to preserve, enhance, and restore environmental quality suitable for life. Its objective is to ensure conditions for socio-economic development, safeguard national security interests, and protect human dignity.

Article 3 establishes relevant concepts for environmental law. For the purposes of Law n. 6.938/81, the environment is defined as the collective conditions, regulations, influences, and interactions of physical, chemical, and biological elements that enable, sustain, and govern life in all its forms (I). The *degradation of environmental quality* is defined as an adverse change in the characteristics of the environment (II). *Pollution* is considered as the degradation of the environmental quality resulting from activities that directly or indirectly: (a) harm the health, safety, and well-being of the population; (b) create adverse conditions to social and economic activities; (c) unfavorably affect the biota; (d) affect the aesthetic or sanitary conditions of the environment; (e) release materials or energy in disagreement with the environmental standards (III). A polluter is the individual, legal entity, or public authority, directly or indirectly responsible for an activity that causes environmental degradation (IV).

[1] Milaré (2018), p. 919.

R. Martins Costa Moreira, G. Wedy, *Brazilian Environmental and Climate Change Law*, https://doi.org/10.1007/978-3-031-87998-2_7

The *environmental resources* are understood as the atmosphere, the inland, surface, and underground waters, the estuaries, the territorial sea, the soil, the subsoil, the elements of the biosphere, the fauna and flora (V).

Law n. 6.938/81 contemplates several rules of great importance for the formulation and implementation of environmental law, such as: the definition of the entities and branches that are part of *SISNAMA* (*Sistema Nacional do Meio Ambiente* or *National System for the Environment*) (Art. 6); the functions of the *CONAMA* (*Conselho Nacional do Meio Ambiente* or *National Council for the Environment*) (Art. 8); the regulatory tools of National Environmental Policy (Art. 9); the regulation of environmental easement (*servidão ambiental*)[2] (Art. 9-A to 9-C); the requirement of environmental licensing for the building, installation, expansion and operation of facilities and activities that use environmental resources, that are effectively or potentially polluting or capable, in any way, of causing environmental degradation (Art. 10); the duty of financial institutions and governmental entities to condition the concession of financing and state incentives to the prior approval of an environmental license and to compliance with environmental regulations (Art. 12); the law establishes provisions for both administrative and criminal liability for environmental violations (Art. 14, §§ 2 to 5; and Art. 15) while maintaining that individuals remain liable to compensate for any harm caused, regardless of intent or fault (strict liability) (Art. 14, § 1); the creation of the Environmental Control and Inspection Fee (*Taxa de Controle e Fiscalização Ambiental* or *TCFA*) (Art. 17-B to 17-H, and 17-P to 17-Q), among others.

Law n. 6.938/81, thus, represents the early landmark of Brazilian environmental law, providing the general normative scope regarding the legal protection of the environment.[3] In the next chapter, we shall study a specific and important part of the National Environmental Policy: The National Environmental System (*SISNAMA* or *Sistema Nacional do Meio Ambiente*).

7.2 National Environmental System (SISNAMA)

The National Environmental System (hereinafter: SISNAMA (*Sistema Nacional do Meio Ambiente*)) is composed of agencies and entities of the Federal Government, the States, the Federal District, and the Municipalities, responsible for the protection and improvement of environmental quality, according to Art. 6 of Law n. 6.938/81.

SISNAMA is formed by the following public organizations:

---

[2]The environmental easement is used for the case where the owner voluntarily limits the use of all or part of his property to preserve, conserve or restore existing environmental resources. The owner can institute environmental easement on his property only for conservation purposes, or to receive government benefits.

[3]Fensterseifer and Sarlet (2021), p. 495.

(a) a superior body, the *Government Council*, with the function of advising the President of the Republic on the formulation of national policy and governmental directives for the environment and environmental resources (Art. 6, I). This council, however, has not had a concrete role in the formulation of environmental guidelines, a role that has been performed by CONAMA.
(b) the National Council for the Environment (hereinafter: CONAMA (*Conselho Nacional do Meio Ambiente*)) is a consultative and deliberative body tasked with advising, researching, and proposing governmental policy guidelines related to the environment and natural resources. It also has the authority to make decisions within its jurisdiction concerning rules and standards that promote an ecologically balanced environment, essential for maintaining a healthy quality of life (Art. 6, II). CONAMA does not have the function of monitoring and detecting violations, nor of issuing administrative sanctions (*poder de polícia* or *pouvoir de police*). However, it is vested with rulemaking authority.
(c) the *Ministry of Environment* (at the time of the Law n. 6.938/81, *Secretariat of Environment*) as a central body, with the purpose of planning, coordinating, supervising, and controlling, as a federal body, the national policy and governmental guidelines set for the environment (Art. 6, III).
(d) the Brazilian Institute for the Environment and Renewable Natural Resources (hereinafter: IBAMA (*Instituto Brasileiro do Meio Ambiente e dos Recursos Naturais Renováveis*) and the Chico Mendes Institute for Biodiversity Conservation (hereinafter: ICMBio (*Instituto Chico Mendes de Conservação da Biodiversidade*), which are federal agencies, to implement and enforce the law, the governmental policies and guidelines set forth for the environment.

Law n. 6.938/81 also provides for sectional agencies (*órgãos seccionais*) and local agencies (*órgãos locais*), which are, respectively, responsible for environmental law enforcement at the state and municipal level (Art. 6, V and VI).

7.2.1 National Council for the Environment (CONAMA)

Due to its strategic importance within the National System for the Environment (SISNAMA), a specific chapter is dedicated to the study of the National Council for the Environment (CONAMA).

As mentioned above, CONAMA is a consultative and deliberative body established to advise, study, and propose governmental policy guidelines for the environment and natural resources. It also has the authority to promulgate environmental regulations and standards within its jurisdiction.[4] The composition, functioning, and structure of CONAMA are laid out in Art. 4 to 6-B, and Arts. 8 to 11 of Decree 99.274/90, which regulates Law n. 6.938/81.

[4]Law n. 6.938/81, Art. 6, inc. II.

CONAMA's functions do not include the exercise of enforcement and repressive authority, a task assigned to the federal (IBAMA and ICMBio), regional, and local enforcement agencies. However, CONAMA wields a rulemaking power, which still generates some controversy. It is worth highlighting the importance of this rulemaking authority: in the absence of an *Environmental Law Code*, CONAMA's regulations apply to all federative entities, in several areas of environmental law.

First, it is important to point out that the regulations promulgated by CONAMA do not have the force of a statutory law enacted by the National Congress, since there is no delegation of legislative power to administrative bodies in Brazil.[5] This rulemaking authority is indeed bound by the principle of legality (Brazilian Constitution, Art. 5, II), which means it cannot contradict legislation enacted by Parliament or introduce entirely new concepts. However, it is not limited to merely reiterating existing legal provisions. Thus, CONAMA holds power not only to interpret legal concepts but also to set guidelines for the issuance of environmental licenses (Law n. 6.938/81, Art. 8, I) and to establish national standards and regulations for controlling pollution from motor vehicles, aircraft, and vessels (VI). Additionally, it is responsible for developing rules, criteria, and standards related to the control and maintenance of environmental quality, with a focus on the rational use of environmental resources, particularly water resources (VII). It is important to note that the specific characteristics of administrative affairs governed by a particular public agency may necessitate greater flexibility in regulation, in accordance with legal provisions and the unique features of the regulated sector. This is exemplified by the broader scope of lawmaking authority granted to regulatory agencies.[6]

The same is true in the ecological field, where reality shows that the complexity and speed of technological evolution and its inherent risks demand from public agents agility, flexibility, and technical knowledge to perform their constitutional duty to protect and preserve the environment. Hence, by promulgating regulations (called in Brazil *Resolutions*), CONAMA can go into technical details that are unworkable for the National Congress.[7]

In fact, the Superior Court of Justice has already decided that CONAMA is empowered by law to issue regulations aimed at protecting the environment and natural resources, including by setting parameters, definitions, and limits of permanent preservation areas.[8] Furthermore, state and municipal rules are also bound by CONAMA's Resolutions, because in relation to environmental protection, it cannot

---

[5] A discussion of this issue can be found at: Moreira and Wedy (2019), pp. 110–112.

[6] On the need for greater freedom for administrative agencies to regulate technical issues, see: Moreira Neto (2007), p. 214.

[7] In this regard: Moreira and Wedy (2019), pp. 112–113.

[8] Superior Court of Justice, REsp 994.881/SC, Rapporteur Min. Benedito Gonçalves, j. 16 Dec. 2008. In the same sense: Superior Court of Justice, REsp 1462208/SC, Rapporteur Min. Humberto Martins, j. 11 Nov. 2014; AgRg no REsp 1183018/MG, Rapporteur Min. Eliana Calmon, j. 07 Mai 2013.

be said that the local or regional interests prevail, and CONAMA is legally authorized to promulgate regulations aimed at protecting ecological reserves.[9]

In another decision, the Superior Court of Justice has held that CONAMA Resolutions result from the legislative mandate, sometimes categorical, sometimes implicit, e.g., Art. 8 of Law n. 6.938/1981.[10] Likewise, it is acknowledged that the requirement for an environmental impact assessment, even when conducted by a state agency, must be evaluated in accordance with CONAMA's Resolution.[11]

More recently, an intense debate over the federal government's deregulatory policy has been called into question. In this regard, CONAMA has recently withdrawn three Resolutions (Resolution n. 284/01, which regulates the environmental licensing of irrigation projects, and Resolutions n. 302 and 303/02, on permanent preservation areas), which were challenged before the Supreme Court because it would mean excessive deregulation and lack of environmental protection. So, the Supreme Court has held to invalidate the abrogation of these regulations, on the grounds that mere revocation of operational rules that set measurable parameters necessary for compliance with environmental legislation, without replacing or updating them, undermines compliance with the Constitution, current legislation, and international commitments.[12] Therefore, the full range of CONAMA's regulations is of great relevance to environmental law, and CONAMA cannot engage in disproportionate deregulatory policies without sound reason.[13] The debate over the most appropriate regulatory measures persists, not only due to attempts to diminish CONAMA's regulations but also because of upcoming legislative changes that are likely to be implemented soon.

Another debate of great relevance is about the composition of CONAMA and the need for greater social participation in this administrative body. On this subject, the reader can consult Sect. 3.7, on the principle of public participation.

[9] Superior Court of Justice, REsp 194.617/PR, Rapporteur Min. Franciulli Netto, j. 16 Apr. 2002.

[10] Superior Court of Justice, AgRg no REsp 1369492/SP, Rapporteur Min. Herman Benjamin, j. 27 Aug. 2013.

[11] Superior Court of Justice, REsp 1330841/SP, Rapporteur Min. Eliana Calmon, j. 06 Aug. 2013.

[12] Brazilian Supreme Federal Court, ADPF 747, ADPF 748, ADPF 749, Rapporteur Min. Rosa Weber, j. 30 Nov. 2020.

[13] For a critical analysis of the revocation of CONAMA's Resolutions for lack of adequate justification, see: MOREIRA (2020).

7.2.2 Authorities Responsible for Implementing Environmental Legislation

In the SISNAMA structure, Law n. 6.938/81 considers the Brazilian Institute for the Environment and Renewable Natural Resources (IBAMA) and the Chico Mendes[14] Institute for Biodiversity Conservation (ICMBio) as *enforcement bodies* (*órgãos executores*), to implement and enforce the environmental law, according to their respective authority (Art. 8, IV). These agencies are linked to the federal government.

At the federal level, the implementation of the National Environmental Policy is primarily handled by IBAMA and ICMBio. This does not, however, exclude the authority of Port Captaincy officers to issue environmental violation notices and initiate administrative proceedings,[15] nor does it limit the authority of the regulatory agencies themselves.

IBAMA, established by Law n. 7.735/1989, is a federal agency (*autarquia*[16] *federal*) under the Ministry of the Environment. It holds the authority to oversee and protect the environment and is responsible for implementing national environmental policies related to environmental licensing, quality control, authorization for the use of natural resources, and environmental monitoring, supervision, and control, all in accordance with the guidelines issued by the Ministry of the Environment, inter alia.

The ICMBio is another federal agency (also an *autarquia*) under the Ministry of Environment. Created by Law n. 11.516/07, it is responsible for proposing, implementing, managing, protecting, inspecting, and monitoring federally protected areas (conservation units), as well as promoting and carrying out research programs focused on the protection, preservation, and conservation of biodiversity. Additionally, it exercises authority over inspection, monitoring, and law enforcement to safeguard biodiversity.

Law n. 6.938/81 considers sectional agencies (*órgãos seccionais*) the State agencies or entities with the power to execute programs and projects, and to control and inspect activities capable of causing environmental degradation; and as local agencies (*órgãos locais*) the municipal agencies or entities, responsible for the control and inspection of these activities, in their respective jurisdictions.

These entities—federal, state, federal district, and municipal—primarily hold the authority to exercise environmental inspection and monitoring powers, referred to in

---

[14] Francisco Alves Mendes Filho, better known as *Chico Mendes,* was a Brazilian rubber tapper, trade unionist, and political activist. He fought on behalf of the rubber tappers of the Amazon Basin, whose livelihood depended on the preservation of the forest and native rubber trees. His activism brought him international recognition, while it provoked the wrath of the local large landowners who assassinated him. His death was reported internationally and provoked outrage in Brazil and abroad. The ICMBio was then created in his honor.

[15] Law n. 9.605/98, Art. 70, § 1.

[16] *Autarquia*, in Brazilian law, is an administrative entity or agency, endowed with its own legal personality, assets and structure, created by law, to provide specific public services.

Brazilian law as *administrative police power* (*poder de polícia*) and in French law as *pouvoir de police*. They achieve these tasks through the application and enforcement of laws, environmental regulations, and CONAMA's standards.[17]

7.3 The Role of other Public Administrative Bodies and Regulatory Agencies

The duty to regulate and enforce environmental legislation is not limited to the bodies and entities mentioned above. Other administrative agencies and entities, including regulatory agencies, can perform this function. In Brazil, we do not have a specific regulatory agency for environmental issues akin to the Environmental Protection Agency (EPA) in the United States. Instead, the responsibility for environmental regulation is distributed across various entities at the federal, state, district, and municipal levels of government.[18]

According to the Superior Court of Justice, Art. 6 of Law n. 6938/81 comprises an open, exemplificative (*numerus apertus*) list of agencies and bodies that are part of SISNAMA, rather than a closed list. This means that it also includes other entities empowered by law to implement environmental regulation, such as the Public Prosecution Service and specialized administrative agencies. In this case, for instance, the Superior Court of Justice held that a fine imposed by the National Petroleum Agency[19] on the company *Petrobrás*[20] for operational failure harmful to the environment, based on Art. 3, IX of Law n. 9847/99, is legally valid.[21]

Another example can be highlighted by the duties assigned to the National Mining Agency (*Agência Nacional de Mineração* or ANM), which, pursuant to Art. 2, XXII of Law n. 13.575/17, is responsible for promulgating regulations and conducting inspections—on a supplementary basis—related to environmental control, hygiene, and safety in mining activities. This agency operates in coordination with other entities tasked with environmental matters and for the hygiene, safety, and occupational health of workers.

[17] Moreira and Wedy (2019), p. 108.

[18] On the role of regulatory agencies in environmental protection in Brazil, see: FREITAS (2014); Moreira and Wedy (2019), pp. 109–110.

[19] This agency, governed by Law n. 9.847/97, is nowadays called *National Agency of Petroleum, Natural Gas, and Biofuels* (*Agência Nacional do Petróleo, Gás Natural e Biocombustíveis* or ANP).

[20] *Petrobrás* is a state-owned company that operates in the areas of exploration, production, refining, commercialization, and transport of oil, natural gas, and their derivatives.

[21] Superior Court of Justice, REsp 1142377/RJ, Rapporteur Min. Herman Benjamin, j. 18 Mar. 2010.

References

Fensterseifer T, Sarlet IW (2021) Curso de Direito Ambiental, 2nd edn. Forense, Rio de Janeiro

Freitas J (2014) Direito Fundamental à Boa Administração Pública, 3rd edn. Malheiros, São Paulo

Milaré É (2018) *Direito do Ambiente*, 11th edn. Revista dos Tribunais, São Paulo

Moreira Neto D d F (2007) Mutações do Direito Administrativo, 3rd edn. Renovar, Rio de Janeiro

Moreira RMC, Wedy G (2019) Manual de Direito Ambiental: de acordo com a jurisprudência dos Tribunais Superiores. Forum, Belo Horizonte

Moreira RMCM (2020) A proteção das APPs e os limites da desregulação nas revogações do Conama. Consultor Jurídico – Conjur, Oct. 6th 2020. https://www.conjur.com.br/2020-out-06/rafael-moreira-revogacao-resolucoes-302-303-conama. Accessed 18 Nov. 2024

Chapter 8
Administrative Tools for Environmental Regulation

8.1 Introduction

The duty of protecting the environment is considered an administrative matter, as it involves the implementation of public policy, which falls under the jurisdiction of the Executive branch. On this basis, the Supreme Court ruled, for instance, on the unconstitutionality of the following laws due to a violation of the principle of separation of powers: (a) a state law that mandated prior approval from the Legislative branch for the preparation of environmental cooperation instruments; and (b) a provision in the state constitution that required licensing and environmental impact assessments to receive prior consent from the Legislative branch.[1]

The Executive Branch, along with environmental administrative agencies and entities, employs various methods to regulate the use of natural resources and oversee activities and facilities that impact the environment. A range of regulatory tools is utilized, including command-and-control measures involving orders, enforcement, and penalties; consensual approaches; and inducement mechanisms such as market tools, incentives, compliance strategies, subsidies, environmental taxation, payment for environmental services, and transparency measures. These instruments can be applied proactively, concurrently, or retrospectively, depending on the stage of the regulated activity.

It is worth noting that in Brazil, under the influence of the *civil law* tradition, particularly French administrative law, the concrete actions of public administration—especially the enforcement of legislation through orders and penalties, or *command-and-control* measures—are generally regarded as an exercise of the so-called *administrative police power*. Nevertheless, the Brazilian legal system has been gradually aligning itself with the concept of regulatory law, as seen in the United States.

[1] Brazilian Supreme Federal Court, ADI 3252 MC, Rapporteur Min. Gilmar Mendes, j. 06 Apr. 2005; ADI 1505, Rapporteur Min. Eros Grau, j. 24 Nov. 2004.

R. Martins Costa Moreira, G. Wedy, *Brazilian Environmental and Climate Change Law*, https://doi.org/10.1007/978-3-031-87998-2_8

In the following sections, we will explore in greater detail the powers of inspection and monitoring, along with command-and-control tools, including the administrative enforcement of environmental law, administrative infractions, and penalties. We will also examine other preemptive control and monitoring measures, such as licensing, environmental impact assessments, and land use regulation, as well as economic instruments like environmental taxation and payment for environmental services.

8.2 Administrative Enforcement of Environmental Law: Administrative Violations and Penalties

When addressing administrative liability, we refer to the duties assigned to administrative bodies and agencies, which are connected to the Executive branch, to define administrative infractions and penalties, as well as to investigate violations and impose penalties on those who violate the law. This administrative function is based on Art. 225, § 3 of the Brazilian Constitution, which sets forth the possibility of imposing on the violator triple liability for the same fact, each with its own requirements: criminal, civil, and administrative. Furthermore, Art. 9, IX of Law n. 6.938/81 includes, as one of the tools of the National Environmental Policy (or a regulatory tool), the "disciplinary or compensatory penalties for non-compliance with the measures required for environmental preservation and recovery of degradation."

The inquiry, processing, and punishment of environmental law violations represent the traditional approach of the environmental administrative *police power* (*poder de polícia*): the repressive or *command-and-control* perspective. Despite the gradual increase in the use of consensual means for the performance of public functions, methods of "command-and-control" do not lose their place of prominence in many situations to prevent and punish violations of the legislation.

Administrative liability for violations of environmental law is primarily governed by Law n. 9.605/98, which sets out general rules regarding the imposition of sanctions. It is up to the federative entities to enact more detailed rules on administrative liability. At the federal level, Decree n. 6514/08 is applicable, outlining the rules governing administrative procedures for environmental sanctions. The states, federal district, and municipalities, in turn, have their own laws on the subject.

8.2.1 Legal Regime of Administrative Liability

Environmental law literature is controversial about the legal regime of administrative environmental liability, i.e., whether it is: (a) subjective liability, depending on the violator's intent, negligence, or recklessness; (b) objective liability (or strict liability), for which the violator's intent is indifferent; (c) or whether it follows a hybrid and peculiar system.

For the first trend, it is a prerequisite for the recognition of administrative liability that the offender has acted with intention, negligence, or recklessness. This is because there could be no crime without willfulness, and because criminal and administrative infractions and sanctions are substantially similar.[2]

The second position holds that administrative environmental liability is solely objective (strict liability), independent of malice or fault on the part of the violator.[3]

For the third line of thought, administrative environmental liability involves a hybrid system between objective (strict) civil liability (that takes place in the case of unlawful behavior) and criminal liability (which, in this case, requires intent or fault). For this line of thought, liability for environmental administrative violations requires intent or fault on the part of the violator. However, there is a *presumption of fault*, so the violator must prove that he or she did not act with intent, negligence, or recklessness.[4]

The Superior Court of Justice, after some disagreement, adopted the thesis that environmental administrative liability should not be considered a matter of strict liability, and depends on *mens rea*. Indeed, according to the Court, the principles of *culpability* (*culpabilidade*) and *non-transcendence of penalties* (*intranscendência das penas*) apply to administrative sanctioning law, in the same way that occurs in criminal law. To better explain this point of view, the Superior Court has ruled that the following individuals or companies do not hold administrative liability: (a) the company that owns the cargo and hires the transporter, in case of an oil spill due to a railroad accident;[5] (b) the purchaser of property in an irregular housing estate for violation of environmental rules committed by the seller;[6] (c) the son who inherits property from his father, due to environmental violation caused by the latter;[7] (d) the owner of a cargo of methanol and fuel oil for environmental damage caused by a ship explosion and the subsequent product leakage, for having only contracted the ship for transportation.[8]

---

[2] Vitta (2003), p. 44; Vitta (2008); Osório (2005), pp. 425–438.

[3] Machado (2024), pp. 350–352. For Freitas and Freitas, the general rule in environmental administrative liability, according to Art. 70 of Law n. 9.605/98, is strict liability. However, only the sanction of *single fine* (*multa simples*) of Art. 72, § 3, II of Law n. 9.605/98 would require fault or intention for its application (Freitas and Freitas (2014), p. 111).

[4] Milaré (2018), pp. 348–362.

[5] Superior Court of Justice, EREsp 1318051/RJ, j. 08 Mai 2019.

[6] Superior Court of Justice, AgInt no AREsp 826046/SC, j. 27 Feb. 2018.

[7] Superior Court of Justice, REsp 1251697/PR, j. 12 Apr. 2012.

[8] Superior Court of Justice, REsp 1401500/PR, j. 16 Aug. 2016.

8.2.2 The Principle of Legality

A central requirement of the rule of law is that public officials should be bound by law. No administrative authority has the discretionary power to violate the law or suspend it.[9] This is the core of the principle of legality, which certainly governs the administrative activity of inspecting and imposing administrative penalties for violations.[10] Indeed, under the Brazilian constitutional system, every act of public administration must be based on law. The public administration can neither contradict the law enacted by the Legislative nor innovate in the legal order, i.e., it cannot create obligations or restrict rights; it must be bound to statutory law.[11] Thus, in the field of administrative law, the government is only allowed to do what statutory law authorizes, and it cannot act upon an inexistent statute.

Therefore, no punitive action can be taken by public officials unless it is authorized by law, as the imposition of administrative penalties on private individuals affects their activities, property, and assets, thereby restricting individual rights. Strict legality requires that both the type of administrative violation and the corresponding penalty are provided for by law passed by the National Congress.[12]

Given this principle, the Supreme Court and the Superior Court of Justice have held that IBAMA[13] is prohibited from applying sanctions provided solely in its internal regulations (*Portarias*).[14]

The fact that the delineation of administrative offenses and penalties must comply with the legality principle is unquestionable. The problem, however, is: What "degree of legality" is required for the definition of an administrative violation? What level of detail is required from the lawmakers, and how much freedom does the Executive retain in specifying a violation and its respective penalty?

This discussion arose mainly due to the open-texture wording of Art. 70 of Law n. 9.605/98, by which: "It is considered an environmental administrative violation any action or omission that violates the legal rules of use, enjoyment, promotion, protection and recovery of the environment." This rule defines administrative environmental violation broadly and generically, as an *open type of violation*, and raises questions about separation of powers and the creation of law by the public administration, without support in legislation enacted by the National Congress. At the

[9] Endicott (2018), p. 22.

[10] Mello (2016), p. 102.

[11] According to Art. 5, II of the Brazilian Constitution, which states that "no one shall be obliged to do or refrain from doing something except by virtue of law," and as per Srt. 37, for which the governmental bodies and entities shall obey, among other principles, the principle of legality.

[12] Garcia et al. (2013).

[13] About IBAMA, read Sect. 7.2.2.

[14] Brazilian Supreme Court, ADI 1823 MC, Rapporteur Min. Ilmar Galvão, j. 30 Apr. 1998; Superior Court of Justice, AgRg no REsp 1.144.604/MG, Rapporteur Min. Hamilton Carvalhido, j. 20 Mai 2010, REsp 1.050.381/PA, Rapporteur Min. Eliana Calmon, j. 16 Dec. 2008; AgRg no REsp 1164140/MG, Rapporteur Min. Humberto Martins, j. 13 Sept. 2011.

federal level, the administrative infractions and penalties are currently specified thereunder Decree n. 6514/2008 (Arts. 24 to 93), promulgated by the President of the Republic, which restates some criminal offenses of Law n. 9605/98 and provides for other conducts. The states, the federal district, and municipalities have also brought their own regulations on the subject. One should remember that the matter is subject to the common and concurrent competence amid the three levels of the federation, as aforementioned.[15]

Despite the controversy, the Superior Court of Justice has held that the administrative rule, combined with the provisions of Art. 70 of Law n. 9.605/98, provides the legal support necessary for the imposition of administrative penalties, and that there is no violation of the principle of legality.[16] Thus, nothing prevents the lawmaker from merely establishing the generic behavior (or generic type) considered illegal, as well as the list and limits of the penalties provided, leaving the specification of these wrongdoings and respective consequences to administrative regulation, by means of a Decree.[17]

The rapid technological evolution, the dynamic nature of environmental balance, and the complexity of challenges in a risk society require that the government and its agencies be granted some flexibility to regulate behaviors that impact the environment and to define the corresponding administrative violations.[18] As mentioned before, technical bodies and agencies may go into details that to the ordinary legislator is not feasible.[19]

In this regard, the same Superior Court decided that the environmental criminal or administrative penalties may refer both to the damage incurred (= result of the degrading conduct) and, alternatively or cumulatively, to the violation of technical requirements for the exercise of the activity or the operational procedure of the enterprise (= *iter* or path of the degrading behavior).[20]

8.2.3 Liable Party

The identification of the party responsible for an environmental violation aligns with the definition of "polluter" under Art. 3, IV of Law n. 6.938/81, which refers to "any

---

[15] Sections 6.2 and 6.3.3.

[16] Superior Court of Justice, REsp 1080613/PR, Rapporteur Min. Denise Arruda, j. 23 Jun. 2009.

[17] Superior Court of Justice, REsp 1137314/MG, Rapporteur Min. Herman Benjamin, j. 17 Nov. 2009.

[18] On the impossibility of a strict and closed definition of unlawful behavior in environmental law, see: Dino Neto (2011); Moreira and Wedy (2019), p. 219.

[19] Section 7.2.1.

[20] Superior Court of Justice, REsp 1142377/RJ, Rapporteur Min. Herman Benjamin, j. 18 Mar. 2010.

individual or legal entity, whether public or private, that is directly or indirectly accountable for an activity causing environmental degradation."

It is worth mentioning that the environmental body or agency can impose a fine on another branch of the government, regardless of the federative entity to which it belongs, against the practice of environmental violation, because there is no hierarchy among the federal, state, federal district, and municipal government. Thus, for example, a municipal environmental agency can issue an environmental violation notice against a federal body.

The application of administrative penalties is subject to the *principle of non-transcendence*, so that, unlike civil liability, it cannot be charged against a third party other than the violator himself. The Superior Court of Justice states that is inherently broader than administrative or criminal liability. Unlike the latter, which does not allow for strict liability of third parties for offenses committed by others, civil liability can encompass a wider range of responsible actors.[21] Accordingly, e.g., the *propter rem* nature of the environmental obligation[22] does not apply to fines and other administrative penalties, but only to civil liability.

8.2.4 Enforcement Authorities

As previously mentioned,[23] according to Art. 23 of the Constitution, all the federative entities have the power to enforce environmental law and there is no exclusivity of one entity or agency to monitor, detect, and punish environmental violations. According to Art. 70, § 1 of Law n. 9605/98, the administrative authorities empowered to issue environmental violation notices and initiate administrative proceedings are the public agents of the environmental agencies that are members of the SISNAMA (National Environmental System), designated for inspection activities, as well as the agents of the Port Captaincies, linked to the Navy, now part of the Ministry of Defense.

At the federal level, this function is mainly ascribed to IBAMA, ICMBio[24] (in relation to violations occurring in federal conservation units) and the Port Authority (within the strict scope of its jurisdiction), without ruling out the authority of the regulatory agencies to impose fines against environmental violations in a supplementary manner. As previously mentioned, the states, the federal district, and

[21] Superior Court of Justice, REsp 1251697/PR, Rapporteur Min. Mauro Campbell Marques, j. 12 Apr. 2012; EREsp 218.781/PR, Rapporteur Min. Herman Benjamin, j. 09 Dec. 2009.

[22] The *propter rem* obligation of civil liability for environmental damage means that the obligation to restore the environment is transmitted to the acquirer or successor of a property, even if the degradation was carried out by the previous owner or possessor.

[23] Section 6.2.

[24] About IBAMA and ICMBio, read Sect. 7.2.2.

municipalities are also empowered to conduct monitoring activities and impose environmental penalties.

Article 17 of Supplementary Law n. 140/2011 gives priority to the agency responsible for licensing or authorizing a project to issue notices of environmental infractions and initiate administrative proceedings to investigate violations of environmental legislation committed by licensed individuals or legal entities. However, such priority does not preclude the common authority of all the federative entities from exercising their enforcement power. Moreover, the Supreme Court has recently ruled that the primacy of the infraction notice issued by the agency originally responsible for environmental licensing or authorization does not exclude the supplementary action of another member of the Federation, provided that an omission or failure to ensure adequate environmental protection is demonstrated.[25]

8.2.5 *Duty of Surveillance by the Public Authorities*

The government's binding obligation to uphold the fundamental right to a healthy environment, together with the principle of proportionality, particularly in the context of preventing insufficient protection (or the state's duty to intervene in environmental matters), leads us to conclude that monitoring, detecting, and penalizing environmental violations by the relevant authorities is mandatory. This commitment falls outside the realm of administrative discretion.

Along these lines, Art. 225 of the Constitution states plainly that the government and the community have the duty to defend and preserve the environment for present and future generations. And § 1 of Art. 225 assigns the public authorities several duties necessary to ensure the effectiveness of environmental law. This idea is reinforced by Art. 70, § 3 of Law n. 9.605/98, which establishes that any environmental authority notified of an environmental violation is required to initiate an immediate investigation through appropriate administrative procedures. Failure to do so renders the authority jointly liable for the infraction.

The Superior Court of Justice has asserted the "unfeasibility of ubiquitous surveillance," the impossibility of imposing on the government the "position of a universal insurer of the entirety of injuries suffered by protected persons or goods," or the unreasonableness of requiring the administration, "for lacking omnipresence," to "prevent any and all acts of lawbreaking," On the other hand, however, the Superior Court has also highlighted that "it is the state's duty to effectively and, in good faith, enforce the rules," being necessary to hold the government accountable, in certain situations, for the omission, remarkably when not performed, in due time, the duty of administrative demolition or other typical acts of self-execution inherent

[25] Brazilian Supreme Court, ADI 4757/DF, j. 12 Dec. 2022.

to the administrative authority.[26] In another decision, this Court ruled that it is the duty of all federative entities to exercise the enforcement authority in the face of an environmental violation. In case of undue omission, they will be liable for environmental damage, jointly with the direct perpetrator of the harm.[27]

Thus, under Brazilian law, public authorities are obligated to monitor, detect, and impose administrative penalties for environmental violations. Should the government fail to fulfill this duty, it may be held accountable for any environmental damage that results from its inaction. Moreover, judges can order the public administration to implement measures to promote the prevention of environmental harm and the preservation of ecosystems.

8.2.6 Administrative Penalties Against Environmental Violations

The range of administrative penalties is specified under Art. 72 of Law n. 9.605/98 (with similar wording in Art. 3 of Decree n. 6.514/08), namely: I—warning; II—simple fine; III—daily fine; IV—seizure of animals, products and byproducts of fauna and flora, tools, equipment or vehicles of any nature used in the infraction; V—product destruction or disablement; VI—suspension of the sale and manufacture of a product; VII—work or activity embargo; VIII—construction or building demolition; IX—partial or total suspension of activities; XI—rights restriction.

If the wrongdoer commits, simultaneously, two or more infractions, the penalties imposed will be applied cumulatively (Law n. 9.605/98, Art. 72, § 1). This list is exhaustive for the federal government unless another penalty is provided by a specific statute, and the states, the federal district, and the municipalities can create other administrative sanctions within their jurisdiction.

The provisions governing the application of these penalties, along with the circumstances under which they may be imposed, are outlined in Art. 72 of Law n. 9.605/98 and, at the federal level, further detailed in Decree 6.514/08. Within their territorial boundaries, states, the federal district, and the municipalities can issue laws and regulations on environmental administrative violations and penalties.

In addition, the penalty imposed, and its amount must be proportional to the seriousness of the fact, taking into account the reasons for the violation and its consequences for public health and the environment, the violator's record of compliance with environmental legislation, and the violator's economic situation (in the federal level, according to Art. 4 of Decree n. 6.514/08).

---

[26] Superior Court of Justice, REsp 1376199/SP, Rapporteur Min. Herman Benjamin, j. 19 Aug. 2014; REsp 1071741/SP, Rapporteur Min. Herman Benjamin, j. 24 Mar. 2009.

[27] Superior Court of Justice, 2ª T., AgRg no REsp 1417023/PR, Rapporteur Min. Humberto Martins, j. 18 Aug. 2015.

Finally, administrative bodies and agencies can still take preventive measures when they detect a violation, such as: I—seizure; II—embargo of work or activity; III—suspension of the sale and manufacture of a product; IV—partial or total suspension of the activity; V—destruction or misutilization of the products, by-products, and instruments of the infraction; VI—demolition. These acts are vested with self-executing authority, carried out by the Public Administration to avoid environmental degradation; notwithstanding, of course, later judicial review, if the violator brings a suit to the Judiciary.

However, the demolition of an inhabited house, whether as a preventive measure[28] or as a final decision by the administrative authority, requires a prior court decision, according to the Superior Court jurisprudence, because the 1988 Constitution requires a previous judicial order (warrant) for a home raid in these cases (Constitution, Art. 5, XI[29]).

Judicial review of administrative decisions is widely available and is based on the fundamental right to effective judicial protection (Constitution, Art. 5, XXXV). Brazil has always adopted, since the beginning of the Republic in 1889, the so-called *English System* or *uniqueness of jurisdiction* (different from the *French System* of administrative jurisdiction), so that the Judiciary has the power to decide upon every dispute, even if it is in the government's interest, being subject to private or public law. Thus, for example, the lack of motivation, the practice of an illegal act or abuse of power, the violation of due legal process, or even the disproportionality of the penalty imposed can be challenged in court by the alleged violator, as widely accepted by Brazilian courts.[30]

8.2.7 Due Administrative Process, Time Limitations, and Judicial Review

The environmental administrative procedure initiated to investigate environmental violations and potentially impose appropriate administrative penalties must, of course, be conducted within the bounds of due process (1988 Constitution, Art. 5, LIV). This ensures the right to adversarial proceedings and a full defense

[28] Article 112, § 3 of the Decree n. 6.514/08 states that demolition as a preventive measure will not be carried out in residential buildings. Nevertheless, the same is true for punishments imposed by the responsible administrative authority at the end of the administrative procedure: Superior Court of Justice, REsp 1217234/PB; REsp 789.640/PB.

[29] Article 5, XI of the Constitution establishes that: "the home is the inviolable refuge of the individual, and no one may enter therein without the consent of the dweller, *except* in the event of flagrant crime (*flagrante delicto*) or disaster, or to give help, or, during the day, by court order."

[30] In this regard, according to several Superior Court of Justice decisions: REsp 462.732/PR, Rapporteur Min. Herman Benjamin, j. 20 Aug. 2009; REsp 1563947/PB, Rapporteur Min. Herman Benjamin, j. 13 Dec. 2016; AgRg no REsp 1048353/SP, Rapporteur Min. Humberto Martins, j. 19 Oct. 2010.

(Constitution, Art. 5, LV; Law No. 9.605/98, Art. 70, § 4) while strictly prohibiting the use of evidence obtained through unlawful means (Constitution, Art. 5, LVI). At the federal level, the Administrative Procedure Act (Law n. 9.784/99) is applied, which grants the opportunity for defense, contradictory, participation, communication, and the right to submit evidence. Furthermore, it is mandatory to provide adequate justification for any administrative decision, in accordance with Art. 50 of Law n. 9.784/99; failure to do so may result in the decision being declared null and void. Note that the states, the federal district, and the municipalities have the power to create their own rules in these regards, notwithstanding the supplementary application of Federal Law n. 9.605/98 and Federal Law n. 9.784/99.

Once an environmental violation is identified, the responsible authority is obligated to take action—both preventively and repressively—to prevent further harm and/or enforce penalties for the breach of environmental legislation. However, this action is subject to time constraints (statute of limitations—*prazo prescricional* or *prescrição*).

The statute of limitations for punitive administrative actions at the federal level is governed by Law n. 9873/99. The time limit for administrative bodies and agencies to undertake any action aimed at initiating an administrative proceeding and enforcing a final penalty is five years, as of the date the act is committed or, in the case of permanent or continuous infringement, from the day it ceases.[31] If the administrative procedure is paralyzed for more than three years, pending judgment or administrative order, the process must be terminated, without preclusion to the assessment of functional liability arising from the undue paralysis.[32] When the fact that is object of the administrative punitive action also constitutes a crime, the statute of limitations will be governed by the time limit provided for under the criminal law[33] (Criminal Code, Art. 109).

For the issuance of a fine, a period of five years is also set for the government to file a special lawsuit (called *fiscal execution* or *execução fiscal*, regulated by Law n. 6.830/80) to collect this fine.[34] One must bear in mind that, in Brazil, the public administration does not possess the authority to carry out expropriatory acts to recover monetary amounts. Instead, the government must turn to the Judiciary to collect any sums owed to it.

---

[31] Law n. 9873/99, Art. 1.

[32] Law n. 9873/99, Art. 1, § 1.

[33] Law n. 9873/99, Art. 1, § 2.

[34] Law n. 9873/99, Art. 1-A.

8.3 Environmental Licensing and Environmental Impact Assessment

Environmental licensing plays a crucial role in environmental regulation in Brazil, as, in principle, any activity or facility that exploits natural resources or has the potential to cause environmental degradation must obtain a prior environmental license before its installation and operation. It stands as one of the most vital tools for environmental regulation and management, owing to its preventive and precautionary nature. It also materializes the polluter-pays and sustainability principles, as well as the socio-environmental function of property.

The environmental licensing process is carried out by an administrative agency or branch belonging to one of the three entities of the Federation. Notably, the Supreme Court has ruled that licensing is an appropriate measure within the administrative authority and falls under the administrative police power, a function of the Executive branch. Consequently, the Court has declared unconstitutional any state laws requiring that licensing and environmental impact assessments be pre-approved by the state Legislature.[35]

Such procedure results mainly from the constitutional duty, imposed on the government, to control the production, commercialization, and use of techniques, methods, and substances that involve risk to life, the quality of life, and the environment (Constitution, Art. 225, § 1, V). This constitutional provision mandates not only the prior assessment of a project's environmental feasibility and sustainability, but also ongoing follow-up and monitoring of the activity or facility throughout its operation, including during the decommissioning and dismantling stages.

The current legal framework of environmental licensing is found in sparse legislation, in CONAMA's Resolutions,[36] and in the state, federal district, and municipal laws. There is not yet a statute in effect on environmental licensing with national or federal application. However, there is currently a bill on environmental licensing under consideration in the National Congress, which has already been approved by the House of Representatives (Lower Chamber) and is awaiting a vote in the Senate. The bill aims to establish general guidelines for environmental licensing across all levels of the Federation.[37]

Licensing is still regulated, partially, by Decree n. 99.274/90, Arts. 16 to 22; Law n. 6938/81, Art. 10; and, regarding the allocation of competence among the federative entities, by Supplementary Law n. 140/11.

[35] Brazilian Supreme Federal Court, ADI 4272, Rapporteur Min. Luiz Fux, j. 30 Aug. 2019; ADI 3252 MC, Rapporteur Min. Gilmar Mendes, j. 06 Apr. 2005; ADI 1505, Rapporteur Min. Eros Grau, j. 24 Nov. 2004.

[36] To read more about CONAMA and its Resolutions, see Sect. 7.2.1.

[37] In the Senate, the bill received the number PL (*Projeto de Lei* or, in literal translation, *Law Project*) n. 2159/21.

As written above, CONAMA's Resolutions[38] are binding on subnational entities (states, federal districts, and municipalities). Thus, for the purposes of environmental licensing and environmental impact assessment, all levels of the Federation must comply with the rules set forth under the following CONAMA's Resolutions: Resolution n. 01/86 and 09/87 on environmental impact assessment; and Regulation n. 237/97 on environmental licensing, among others that deal with specific issues.

It is important to differentiate between two concepts: a) *environmental licensing* (*licenciamento ambiental*) is an administrative procedure that results in a decision on whether to grant or deny a license for activities or projects that involve the use of environmental resources or have the potential to cause environmental degradation (Supplementary Law n. 140/11, Art. 2, I); and b) *environmental license* (*licença ambiental*) is the final administrative act by which the responsible environmental agency or body sets the conditions, restrictions and environmental control measures that must be complied with by the applicant (CONAMA Regulation n. 237/97, Art. 1, II).

Environmental licensing does not exclude nor substitute other allowances or permits necessary for a certain activity to be implemented and executed. Article 10, § 1 of CONAMA Resolution n. 237/97 stipulates that, in addition to ensuring that the project complies with land use and zoning regulations under municipal law, it must also obtain, where applicable, authorization from the competent authorities for vegetation clearance and water usage. It is important to highlight that, since the enactment of Supplementary Law 140/11, the authorization for vegetation suppression as part of environmental licensing is granted by the same body or agency responsible for issuing the license (Art. 13, § 2).

Thus, the granting of an environmental license does not replace the need to get, for example, the water resources permit from the appropriate governmental authority (according to Art. 12 of Law n. 9433/97), or authorization for the exploitation of mineral resources (Decree-Law n. 227/67, Art. 7; Law n. 13.575/17).

This theme brings us to another issue of great importance: the participation of the so-called *intervening bodies* (*órgãos intervenientes*) or *involved authorities* (*autoridades envolvidas*), that is, those administrative entities or agencies specialized in certain matters related to the activity subject to environmental licensing. This is the case, for example, of agencies responsible for regulating water use,[39] protecting Indians[40] and *quilombolas* (runaway slaves),[41] protecting cultural

[38] See Sect. 7.2.1.

[39] The National Water Agency (*Agência Nacional de Água* or *ANA*) was given the task of implementing the National Water Resources Policy and establishing reference standards for the regulation of public basic sanitation services (Law n. 9.984/2000).

[40] The implementation of the Indigenous protection policy is performed by the National Indian Foundation (*Fundação Nacional do Índio* or *FUNAI*) (Law n. 5.371/67).

[41] The creation of the Palmares Cultural Foundation (*Fundação Cultural Palmares*) was authorized by Law n. 7.668/88, to promote the preservation of the cultural, social, and economic values resulting from the black influence in the formation of Brazilian society. The expression *quilombolas* refers, in short, to the descendants and remainders of communities formed by runaway slaves (the *quilombos*), between the sixteenth century and 1888 (when slavery was abolished), in Brazil.

heritage,[42] managing conservation units, etc. However, the opinion of these intervening agencies, while mandatory, is not binding on the administrative body responsible for the licensing process and must adhere to the established deadlines. Naturally, the agency or body responsible for conducting the licensing and making decisions on the license application must provide sufficient justification for any decision that rejects the opinion of the involved authority.[43]

The exception stands for the cases of environmental licensing for activities and facilities with significant environmental impact that may affect a specific conservation unit or its *buffer zone* (*zona de amortecimento*). In this situation, a previous authorization from the body responsible for managing the conservation unit is required, and its opinion is binding (Law n. 9.985/00, Art. 36, § 3). In other licensing cases that affect conservation units in some way, there are specific rules in CONAMA Resolution n. 428/2010, but in these cases, the advice of the conservation unit's management body is not binding.

The licensing process and the environmental studies ought to be accessible to the public, as a rule, according to the transparency principle. Therefore, pursuant to § 1 of Art. 10 of Law n. 6.938/81, the "request for a license, its renewal, and concession will be published in the official gazette, as well as in a regional or local newspaper of wide circulation, or by means of electronic communication managed by the appropriate environmental agency." The disclosure of the environmental impact assessment and its respective report is required by the Brazilian Constitution, under Art. 225, § 1, IV.

The legislation, however, safeguards commercial, industrial, and financial secrecy, and any other secrecy protected by law, as well as that related to the internal communications of governmental bodies and entities (Decree n. 99.274/90, Art. 17, §§ 3 and 4; Law n. 10.650/03, Art. 2, § 2; CONAMA Resolution n. 01/86, Art. 11).

Finally, it is worth mentioning that environmental licensing raises financial costs—sometimes quite high when the preparation of an environmental impact study is required. These expenses, which are incurred by the license applicant to

[42] At the federal level, the protection of cultural heritage is carried out by the Institute for National Historic and Artistic Heritage (*Instituto do Patrimônio Histórico e Artístico Nacional* or *IPHAN*) (Law n. 378/37). The states, the federal district, and municipalities can create their own bodies to protect the cultural heritage.

[43] This statement results mainly from Art. 13, § 3 of LC n. 140/11, and from Interministerial Ordinance (*Portaria Interministerial*) n. 60/15 (signed by the Ministers of the Environment, Justice, Culture and Health). Similarly, the environmental licensing bill (PL 2159/21) establishes that the opinion of the authorities involved is not binding. In this regard, read: Trennenpohl and Trennenpohl (2020), p. 245.

cover the costs of the licensing process, should not be mistaken for the taxes levied for the exercise of environmental administrative authority.[44]

Article 13 of CONAMA Resolution n. 237/1997 states that the "cost of analysis for obtaining an environmental license shall be established by legal provision, aiming at the reimbursement, by the entrepreneur, of the expenses incurred by the environmental agency." It should be noted that under Art. 17 of this Resolution, such cost must be paid again in the case of a new license application, after the previous one has been dismissed due to non-compliance with the deadline of Art. 15. In addition, Art. 13, § 3 of Supplementary Law n. 140/2011 establishes that the amounts related to *fees* for environmental licensing and other services must be proportional to the cost and complexity of the service provided by the administrative authority.

8.3.1 Activities Subject to Environmental Licensing

Not every economic activity needs an environmental license to be implemented and operated. A license will only be required if the activity is effectively or potentially polluting, or can cause environmental degradation in any form, as clearly outlined in Art. 10, § 1 of Law n. 6.938/81, Art. 1, inc. I of CONAMA Resolution n. 237/1997 and Art. 2, inc. I, of Supplementary Law n. 140/2011.

Therefore, environmental licensing serves as a mechanism for regulating economic activities in light of the fundamental right to a balanced environment. Activities that do not pose a threat to this right have no reason to be subject to such a permit. Consequently, not all economic activities require licensing, but only those with the potential to cause pollution or environmental harm that is more than negligible.[45]

Currently, CONAMA Resolution n. 237/97 lists the activities that are considered to have an environmental impact and for which environmental licensing is required. This is a merely sample list because other activities and projects may also require a previous license to operate if they might cause environmental harm.

The legislation uses, therefore, an *open-texture term* or an *indeterminate legal concept* (*conceito jurídico indeterminado*), that is: "activities or facilities that use environmental resources," "effectively or potentially polluting," or "capable, in any way, of causing environmental degradation." *Open-texture terms* or *indeterminate legal concepts* are those whose content and extension are largely uncertain, that is, they are not endowed with a precise and objective meaning. The application of these

---

[44] The *Environmental Control and Inspection Tax* (*Taxa de Controle e Fiscalização Ambiental* or TCFA) is a kind of tax, based on Art. 145, II of the Constitution, whose taxable event is the regular exercise of the administrative police power (*poder de polícia*) vested in the IBAMA for the control and inspection of potentially polluting activities and users of natural resources. It is regulated by Art. 17-B to 17-H of the National Environmental Policy Act (Law n. 6.938/81).

[45] Farias (2024) pp. 37–51.

concepts grants judges and public officials considerable latitude for interpretation and discretion. There is a controversy among administrative law scholars as to whether these concepts engender administrative discretion or allow for a single interpretation that could be considered the correct one.[46]

In any case, under the Brazilian legal system, the possibility of challenging the decision of an environmental public agency or body regarding the necessity of environmental licensing in a particular case is not in question and may be pursued through legal action. Judicial review takes place both in the case of improper waiver of licensing, as well as to question its grounds and even the decision to grant a license from the standpoint of socio-environmental feasibility or sustainability of a work or activity.

As has been decided by the Superior Court of Justice, in Brazil, it must prevail the principle of the *unavoidability of judicial review* (*inafastabilidade do controle judicial*), because any administrative action is, in principle, reviewable by the Judiciary. Although it is up to the administrative body to decide whether an activity requires a previous license and an environmental impact assessment, this decision can be challenged and quashed by the courts.[47] This does not mean, however, that the Judiciary has become the "environmental agency's branch," making it in charge of originally analyzing licensing applications. It is the judge's role only to check administrative actions, to define parameters for compliance by the government and, if necessary, to demand the enforcement of a court decision.[48]

8.3.2 Federative Competence for Environmental Licensing

The licensing authority allocation among federative entities, despite the efforts in its systematization, especially after the enactment of Supplementary Law n. 140/11, is still a source of conflict and legal uncertainty. This topic will analyze the rules about the division of roles between the federal, federal district, state, and municipal governments to analyze environmental licensing applications and the prevailing understanding of controversial issues.

---

[46] For insights into the controversy surrounding indeterminate legal concepts, the presence of discretion, or the need for interpretation, see: Mello (2010), pp. 22–32; Di Pietro (2024), pp. 224–5; Oliveira (2014), pp. 94–97; Enterría and Fernández (1990), pp. 392–99); Pires (2009), pp. 91–122.

[47] Superior Court of Justice, REsp 1279607/PR, Rapporteur Min. Mauro Campbell Marques, j. 06 Dec. 2011.

[48] Superior Court of Justice, REsp 1807851/SP, Rapporteur Min. Napoleão Nunes Maia Filho, j. 03 Dec. 2019.

8.3.2.1 Common Jurisdiction and Cooperative Federalism

The power to decide on an environmental license application, an expression of the state's duty to protect the environment, is common to all political entities of the federation, pursuant to Art. 23, III, IV, VI, and VII of the Brazilian Constitution. In addition, as stated in Art. 17-L of Law n. 6.938/81, the licensing, registration, authorization, concession, and permission related to fauna, flora, and environmental control are vested with the agencies that are part of the National Environmental System (SISNAMA). This authority, therefore, falls under cooperative federalism. While it increases the number of entities responsible for environmental protection, it also leads to conflicts of jurisdiction, overlaps, and duplication of efforts.[49]

For this reason, the Constitution (Art. 23, sole paragraph) set that Supplementary Laws (*Leis Complementares*) shall establish rules for the cooperation between the federal government and the states, the federal district, and the municipalities, aiming at the attainment of balanced development and well-being on a nationwide scope. Thus, to resolve or mitigate federative conflicts and promote legal certainty,[50] Supplementary Law n. 140/2011 was enacted, which laid down rules for cooperation among national and subnational entities in administrative actions resulting from the exercise of common competence relating to environmental protection and regulation.

Article 4 of Supplementary Law n. 140/11 provides relevant tools of federative cooperation, namely public partnerships, agreements, public boards (commissions), public or private funds, and other economic tools, and delegation of functions and execution of administrative actions.

To ensure legal stability and avoid overlapping actions of federative entities, the legislation adopted the concept of *unicity of licensing* (*unicidade do licenciamento*). In line with this, Art. 13 of Supplementary Law n. 140/11 establishes that projects and activities are to be environmentally licensed or authorized by a single federative entity. Besides, other federative entities may provide their input to the responsible licensing or authorizing agency, though only in an advisory capacity, and must comply with the established timelines and procedures for environmental licensing (Supplementary Law n. 140/11, Art. 13, § 1).

8.3.2.2 Allocation of Authority Under Supplementary Law n. 140/11

The allocation of licensing authority had been defined previously under CONAMA Resolution n. 237/97. Then, Supplementary Law n. 140/11 was enacted with the aim

[49] Farias (2024), pp. 139–152.

[50] On the importance of Supplementary Law n. 140/11 to solve federative conflicts in environmental matters, read: Farias (2020), pp. 3–6.

of rationalizing and clarifying the roles of the federative entities, reducing conflicts and establishing instruments of cooperation.[51]

The Supplementary Law n. 140/11 has changed former rules, provided by CONAMA Resolution n. 237/97 and Law n. 6.938/81, Art. 10 and § 4. The most significant change was the shift from using the geographical extent of direct environmental impacts as the criterion to focusing on the location of the polluting activity itself. The purpose was to promote greater certainty because often the delimitation of direct environmental impacts can only be done in the course of the licensing procedure.[52]

Article 7, XIV of Supplementary Law n. 140/11 grants the Federal Government (*União Federal*) the power to promote the licensing of projects and activities in the following cases: (a) located or developed jointly in Brazil and in a bordering country; (b) located or developed in the territorial sea, continental shelf or exclusive economic zone; (c) located or developed in Indigenous lands; (d) located or developed within Conservation Units established by the Federal Government, except in the "Environmental Protection Areas" (*Áreas de Proteção Ambiental* or APAs[53]); (e) located or developed in two or more states; (f) of a military nature, excepting from environmental licensing, under the terms of an act of the Executive, those envisaged in the preparation and employment of the Armed Forces; (g) those aimed at researching, mining, producing, benefiting, transporting, storing, and disposing of radioactive material, at any stage, or that use nuclear energy in any of its forms and applications, upon opinion from the National Nuclear Energy Commission.

The federal competence for environmental licensing may still be extended in two other situations: (a) according to Art. 7, XIV, "h," for projects that meet the typology established by an act of the Executive, based on a proposal from the National Tripartite Commission, and considering the criteria of size, polluting potential, and the nature of the activity or facility; (b) according to Art. 7, sole paragraph, for projects whose localization comprises jointly areas of the land and sea lanes of the coastal zone, the Federal Government will be responsible exclusively in the cases set in the typology established by an act of the Executive, based on the proposal of the National Tripartite Commission, and considering the criteria of size, polluting potential, and nature of the activity or enterprise. In 2015, Decree n. 8.437 was issued, assigning the Federal Government (more specifically, the federal agency IBAMA[54]) to carry out environmental licensing for other activities, especially involving federal infrastructure works, oil and gas exploitation, inter alia.

---

[51] Farias (2020), pp. 152–153.

[52] Farias (2024), pp. 152–153.

[53] *Área de Proteção Ambiental* (APA) or *Environmental Protection Area* is a sort of Conservation Unit, which normally covers an extensive area, with a certain degree of human occupation, endowed with abiotic, biotic, aesthetic or cultural attributes that are especially important for the quality of life and well-being of human populations, and has as its basic purposes to protect biological diversity, regulate the occupation process and ensure the sustainability of the use of natural resources. It is regulated by Art. 15 of Law n. 9985/2000.

[54] To read more about IBAMA, see Sect. 7.2.2.

Apart from the cases explicitly outlined in legislation (Supplementary Law n. 140/11) or by Decree (Decree n. 8.437/2015), the responsibility for issuing licenses for other activities falls to the states (and the federal district, in matters of state jurisdiction). That is, as set by Art. 8, XIV of Supplementary Law n. 140/11, the states have *residual competence* (*competência residual*), which is why the bulk of environmental licensing in Brazil is conducted at the state level.

It also established the competence of the municipalities to promote the environmental licensing of activities or facilities that cause or may cause local environmental impact, according to the typology defined by the respective state environmental councils, considering the criteria of size, polluting potential, and nature of the activity (Art. 9, XIV of Supplementary Law n. 140/11). Finally, the activities located and developed in Conservation Units created by states and municipalities will be subject to licensing by the unit's conservation creator, except in the case of "Environmental Protection Areas," as mentioned above (Supplementary Law n. 140/11, Art. 8, XV; Art. 9, XIV, "b"). The federal district's competence, it is worth remembering, includes that of the states and municipalities (Supplementary Law n. 140/11, Art. 10).

It should be noted that the delegation of duties by means of an agreement is possible as provided under Art. 5 of Supplementary Law n. 140/11, as long as the delegating entity has a capable body and an environmental council. It is common, therefore, for states to delegate to municipalities the task of conducting licensing procedures for certain activities that have predominantly local impacts or that can be more effectively handled at the municipal level. However, for this to occur, municipalities must have an environmental council and a qualified environmental body.

The suppression of vegetation resulting from environmental licensing must be authorized by the same agency or entity that conducts the licensing process. (Supplementary Law n. 140/11, Art. 13, § 2). It is important to point out that, according to the Forest Protection Code and Law n. 6.938/81, in accordance with the interpretation of the Superior Court of Justice, native flora, in the case of suppression, is uniformly protected by the need for prior and valid authorization from the appropriate environmental agency, whatever its biome, location, typology, or state of conservation.[55] Thus, the suppression of native vegetation outside the context of an environmental license also requires prior authorization, as governed by the Forest Code (Law n. 12.651/12), the Atlantic Forest Act (Law n. 11.428/06), and the Supplementary Law n. 140/11 (Art. 7, XV; Art. 8, XVI; Art. 9, XV, "a," "b").

As noted above, in cases of environmental licensing for activities and facilities with significant environmental impact, which requires the preparation of an environmental impact assessment, that may affect a specific conservation unit or its *buffer zone* (*zona de amortecimento*), authorization is required from the agency or body responsible for managing the unit (Law n. 9.985/00, Art. 36, § 3).

[55] Superior Court of Justice, REsp 1198727/MG, Rapporteur Min. Herman Benjamin, j. 14 Aug. 2012.

Supplementary Law n. 140/11 also deals with the so-called *supplementary competence* (*competência supletiva*) and *subsidiary competence* (*competência subsidiária*). The first is applicable when the municipal, federal district, or state entity does not meet the required deadlines to analyze licensing applications, or when it does not have a capable environmental agency or environmental council. In these cases, the state can assume the municipal functions, and the federal entity can assume the state and federal district functions (Supplementary Law n. 140/11, Art. 2, II and Art. 15). If the federal entity (IBAMA[56]) does not meet the time limit, however, there is no legal provision for states or municipalities to assume federal competence. Under these circumstances, the applicant can only turn to the Judiciary to remedy governmental inertia.

The Supreme Court, in a recent decision, extended the incidence of supplementary competence to two other cases: (a) if the environmental agency delays in analyzing an application for license renewal; and b) for the case of omission or insufficiency in environmental supervision by the licensing agency.[57] In the former case, Art. 14, § 4 of Supplementary Law n. 140/11 provided that the renewal of environmental licenses must be requested at least 120 days prior to the expiration of its validity term, established in the respective license, which is automatically extended until the definitive decision of the proper environmental agency. However, the Supreme Court declared the provision for automatic renewal of environmental licenses unconstitutional. Instead, if the environmental agency delays in deciding on a license renewal application, the supplementary competence rule should be applied. Regarding the latter case, Supplementary Law n. 140/11 stipulates that the agency responsible for environmental licensing holds priority in issuing infraction notices and imposing administrative penalties (Art. 17). Nevertheless, if it is proven that the responsible agency was negligent or inadequate in monitoring environmental violations, supplementary authority may be invoked, allowing another public entity to step in and address the administrative infractions.

Subsidiary competence, in turn, refers to the intervention of one federative entity to assist another in performing specific functions, upon the request of the entity originally responsible for that activity (Supplementary Law n. 140/11, Art. 2, III). It may take place by means of technical, scientific, administrative, or financial support, and other means of cooperation (Supplementary Law n. 140/11, Art. 16).

It is worth bearing in mind that the authority for monitoring environmental violations is common to all entities of the Federation, not limited to the entity with power for licensing. Article 17 of Supplementary Law n. 140/11 establishes that the agency responsible for licensing or authorization, as the case may be, of a facility or activity, has the authority to issue a notice of environmental violation and initiate administrative proceedings for the investigation of violations committed by the licensed or authorized person or legal entity. However, § 2 of Art. 17 brings the first exception, by prescribing that, in cases of imminence or occurrence of

[56] About IBAMA, read Sect. 7.2.2.

[57] Brazilian Supreme Federal Court, ADI 4757/DF, Rapporteur Min. Rosa Weber, j. 12 Dec. 2022.

environmental degradation, the public agent that detects the fact can take measures to avoid it, make it cease, or mitigate it, immediately communicating to the agency responsible for the license to take appropriate measures. Moreover, by § 3 of Art. 17, the performance of the common competence of all the federative entities to inspect the conformity of facilities and activities with the environmental legislation is not prevented, prevailing the act of the agency that holds the licensing or authorization authority. In any case, as ruled by the Supreme Court, if it is demonstrated that there has been omission or inadequacy in the monitoring function, another entity may assume this role under the rule of supplementary competence.

The Superior Court of Justice has clarified that the function of granting licenses should not be confused with the authority to monitor and supervise environmental violations, and there is no question of an exclusive power vested in any specific federative entity to implement protective measures. Hence, the environmental enforcement authority must be exercised by all four federative entities, regardless of the place where the threat or damage is taking place, as well as the licensing competence. For this reason, it has been understood that the inspection of activities that are harmful to the environment grants IBAMA sufficient legal interest to exercise its administrative authority for inspection and punishment of environmental infractions, even if the violation takes place in an area where the licensing competence is held by the municipality or the state.[58]

8.3.3 Types of Environmental Licenses

Pursuant to Art. 19 of Decree n. 99.247/90, and Art. 8 of CONAMA Resolution n. 237/97, there are basically three types of licenses:

(I) The *Preliminary License* (*Licença Prévia*, or LP) is granted in the initial phase of activity planning, establishing the fundamental requirements to be fulfilled during the location, installation, and operation stages, in accordance with municipal, state, or federal land use regulations.

(II) The *Installation License* (*Licença de Instalação*, or LI) authorizes the commencement of construction for a project or facility.

(III) The *Operating License* (*Licença de Operação* or LO) is issued after the necessary inspections have been conducted, authorizing the commencement of the licensed activity and the operation of its pollution control equipment, in accordance with the conditions set forth in the previous licenses.

It is important to note that the issuance of a Preliminary License does not authorize the construction or operation of a project, nor does it result in the automatic granting of subsequent licenses. The interested party has no vested right, but a mere

[58] Superior Court of Justice, AgInt no REsp 1484933/CE, Rapporteur Min. Regina Helena Costa, j. 21 Mar. 2017; AgInt no REsp 1530546/AL, Rapporteur Min. Herman Benjamin, j. 07 Feb. 2017.

expectation to obtain the following licenses (Installation (LI) and Operation (LO) Licenses).

Article 12 of CONAMA Resolution n. 237/1997 states that the environmental agency will define, if necessary, specific procedures for environmental licenses, taking into account the nature, characteristics, and peculiarities of the activity, and also the compatibility of the licensing procedure with the stages of planning, implementation, and operation. This rule provides for three differentiated types of licensing procedures: (a) streamlined license for activities of low environmental impact; (b) single license for multiple neighboring facilities (similar to a pollution bubble concept); and (c) a more flexible and streamlined procedure for environmental licensing for activities and enterprises that implement voluntary environmental management plans and programs, aimed at promoting continuous improvement and enhancing environmental performance.

Regarding streamlined procedures for activities with low environmental impact, there are several CONAMA Resolutions, e.g., electric enterprises with low potential for environmental impact (CONAMA Resolution n. 279/2000); low-impact carciniculture in the coastal zone (CONAMA Resolution n. 312/02); sewage systems (CONAMA Resolution n. 377/06); small agribusinesses with low potential of environmental impact (CONAMA Resolution n. 385/2006); social housing projects, in urban areas or urban expansion areas (CONAMA Resolution n. 412/2009); aquaculture (CONAMA Resolution n. 413/2009); railway projects with low environmental impact and the regularization of the ongoing activities (CONAMA Resolution n. 479/2017).

The possibilities and limitations of simplifying environmental licensing have recently garnered significant attention. Many states have adopted simplified licensing models, such as the so-called *licensing by adherence and commitment* or by *self-declaration* (*licença por adesão e compromisso* (LAC) or *licença por autodeclaração*), where the installation, expansion, and operation of a facility or activity is permitted, provided certain conditions are met. This process typically involves a simple declaration of commitment submitted to the environmental agency, often electronically, without prior inspection or assessment. This type of license may be contemplated in the licensing bill (PL n. 2.159/21), commented above, and the Supreme Court, so far, considered its practice by the states constitutionally valid, because it was adopted within their power to tailor the procedure to regional peculiarities.[59]

In a more recent ruling, the Supreme Court reaffirmed this understanding by upholding the validity of a state law that establishes a simplified environmental licensing model for the regularization of activities or enterprises in the process of installation or operation, as well as for activities with low to medium pollution potential. In this case, a law from the State of Bahia introduced the *Regularization License* and the *Environmental License by Adhesion and Commitment*, which fall

[59] Brazilian Supreme Federal Court, RE 1264738 AgR, Rapporteur Luiz Fux, j. 24 Aug. 2020; ADI 4615, Rapporteur Min. Roberto Barroso, j. 20 Sept. 2019

within the scope of concurrent legislative jurisdiction and fulfill the constitutional duty to supplement legislation on environmental licensing, considering the predominance of local interest in establishing specific procedures for activities and enterprises.[60]

The aforementioned licensing bill also intends to introduce the possibility of a *single license* (*Licença Única*), which consolidates the three phases—Preliminary, Installation, and Operating License—into one comprehensive approval. This procedure, however, was held unconstitutional by the Supreme Court, when analyzing a state rule, on the grounds that such simplification would result in an infringement on the power of the National Congress to enact general rules regarding environmental protection, as well as a violation of the precautionary principle and the constitutional duty to protect the environment.[61]

The Supreme Court has also imposed limits on the streamlining introduced by state laws that exempt certain economic activities from licensing and environmental assessment, as such measures could undermine environmental protection.[62]

Recently, the dispute over environmental license streamlining got a new chapter: the Federal Government, by means of a *Provisional Measure* (*Medida Provisória*[63]), made into Law n. 14.195/21, allowed the automatic concession of an operation permit and license for companies that carry out activities classified as "medium risk," without human analysis. Such a procedure, for example, could remove the prior control of activities that are potentially harmful to the environment, such as the transfer of oil cargoes and derivatives on the high seas, the economic exploitation of timber or firewood and forest by-products and the manufacture of fertilizers and agrochemicals. This provision triggered a lot of criticism and controversy. It ended up leading to a judicial review lawsuit before the Supreme Court. Eventually, the Court ruled that this rule is unconstitutional to automatically grant an environmental license for the operation of companies that carry out activities classified as medium risk, among other measures.[64] Thus, the Supreme Court has been imposing limits on the simplification of the environmental licensing process, by deciding that this streamlining cannot generate an absence of state control over polluting activities.

---

[60] Brazilian Supreme Federal Court, ADI 5014/BA, j. 11 Nov. 2023.

[61] Brazilian Supreme Federal Court, ADI 5475, Rapporteur Min. Cármen Lúcia, j. 20 Apr. 2020; ADI 6672/RR, Rapporteur Min. Alexandre de Moraes, j. 14 Sept. 2021.

[62] For the unconstitutionality of state legislations that waive licensing for particular economic segments, without considering the dimension of the impacts in the individual case, see: STF, ADI 6650/SC, Rapporteur Min. Cármen Lúcia, j. 26 Apr. 2021; ADI 6288, Rapporteur Rosa Weber, j. 23 Nov. 2020.

[63] It is worth remembering that the *Provisional Measure* or *Medida Provisória* is a law adopted by the President of the Republic in important and urgent cases, which shall be submitted to the National Congress immediately to be ratified, and thereby become final law, or rejected. It is governed by Art. 62 of the Brazilian Constitution.

[64] Brazilian Supreme Federal Court, ADI 6808, Rapporteur Min. Carmen Lúcia, j. 28 Apr. 2022.

8.3.4 *Validity Periods of Licenses and Deadlines for Decisions on License Applications*

Environmental licenses are not everlasting but are subject to term limits. Law n. 6.938/81, by providing for the revision of the license (Art. 9, IV) and by mentioning its renewal (Art. 10, § 1), emphasizes this non-eternal character of any license, aiming to prevent the perpetuation of standards that are often technologically outdated.

The expiration period of licenses is currently regulated by Art. 18 of CONAMA Resolution n. 237/97, according to which the responsible environmental agency will establish the expiration period for each type of license, specifying them in the respective document, taking into consideration the following aspects: (I) The expiration term of the Preliminary License must be, at least, the time established by the elaboration schedule of the plans, programs and projects related to the facility or activity, and cannot exceed five years; (II) The period of validity of the Installation License must be, at least, the period established by the installation schedule of the facility or activity, and cannot exceed six years; (III) The expiration term of the Operation License must take into consideration the environmental control plans and will be of at least four years and at most ten years.

The Preliminary License and the Installation License may have their expiration term extended, if they do not exceed the maximum durations established under Art. 18 (CONAMA Resolution n. 237/97, Art. 18, § 1). The environmental agency may establish specific validity periods for the Operating License of facilities or activities that, due to their nature and peculiarities, are subject to closure or modification in shorter periods (§ 2). When renewing the Operating License of an activity, the environmental agency may, by means of a properly reasoned decision, increase or decrease its validity term after evaluating the environmental performance of the activity, respecting the limits established in this Resolution (§ 3). The renewal of the Operation License of an activity or facility shall be requested at least 120 days prior to the validity term established in the respective license. This rule, as well as Art. 14, § 4 of Supplementary Law n. 140/11, stated that the Operation License may be automatically extended until the definitive decision of the environmental agency has been taken. However, as we reported above, the Supreme Court has decided on the unconstitutionality of this automatic renewal. The environmental agency's failure to decide on a license renewal application now results in the supplementary competence of another federative unit to review the request.[65] States, the federal district, and municipalities can establish their own deadlines, but cannot exceed the time limits set under CONAMA Resolution n. 237/97.[66]

---

[65] See Sect. 8.3.2.2.

[66] Milaré is of the opinion that respecting the time limits established by these basic federal guidelines, state and municipal legislators, taking local peculiarities into account, may provide for different periods of licenses validity (Milaré (2018), p. 1107). According to Farias, CONAMA Resolutions carry the weight of a general federal rule, meaning that state and municipal authorities

In relation to the deadlines for analyzing and processing license applications, Art. 14 of Supplementary Law n. 140/2011 stipulates that licensing authorities must adhere to the established timelines for initiating licensing procedures. This provision, which arguably should not even be necessary—given that the obligation to meet deadlines is implicit, with failure to do so constituting unlawful administrative inertia—is not consistently followed in practice. The delay is largely attributed to the fact that many environmental agencies lack sufficient human and material resources to handle the growing volume of administrative tasks.[67]

The expiration of the deadlines for deciding on a licensing application, without the responsible agency issuing the environmental license, neither implies tacit approval nor authorizes the performance of any action that depends on or stems from it. However, it does activate the supplementary competence previously discussed[68] (Supplementary Law n. 140/2011, Art. 14, § 3).

8.3.5 License Amendment and Withdrawal

The license amendment and withdrawal are governed by Art. 19 of CONAMA Resolution n. 237/97. Thus, the license may be withdrawn for violation or inadequacy of any legal conditions or rules, as well as for omission or false description of relevant information that supported its issuance.

In addition, it is also possible to revoke the license, despite its legality, due to unexpected serious environmental and health risks. That is, even if the entrepreneur is acting lawfully and complying with all the rules, this does not guarantee that the license will never be undone. For an environmental authority to revoke a valid license, it must clearly demonstrate that the activity causes intolerable environmental harm, based on new facts or information uncovered by the agency. Therefore, the environmental license is granted *rebus sic stantibus*, meaning it is based on the factual circumstances as they existed at the time of issuance. This characteristic of environmental licensing is largely due to the constantly changing nature of environmental balance, the rapid technological advancements, and the interaction between humans and nature, which require continuous adaptation to changes in the characteristics of environmental impacts and the cost-benefit analysis of the project.[69]

---

are only responsible for supplementing or addressing any gaps in the federal legislation (Farias (2024), pp. 106–110). In our view, the time limits established by CONAMA Resolution n. 237/97 contemplate a minimum level of environmental protection that must be observed by the other entities of the Federation, which is why they cannot extend the validity terms of the licenses beyond those established in this rule.

[67] Moreira and Wedy (2019), p. 148.

[68] Section 8.3.2.2.

[69] Milaré (2018), pp. 1107–1108; Moreira and Wedy (2019), pp. 155–156.

If, on the one hand, there is undeniable the possibility of changing and revoking the environmental license granted, on the other hand, the right to compensation in favor of the entrepreneur in these cases raises some controversy.

If the interested party causes the invalidation of the license or fails to comply with the conditions imposed, compensation will not be awarded if the act is made void.

Some commentators understand that, in view of the application of specific principles of environmental law, notably the polluter-pays and the precautionary principles, as well as in view of the adoption of strict liability for causing environmental damage based on full risk (as will be analyzed below), one cannot speak about indemnification to the holder of a revoked license.[70]

However, the prevailing understanding is that in case of license cancellation without fault of the entrepreneur or unlawful behavior, compensation is due to those who made investments and trusted in the validity of the public administration's decision. This is the most appropriate approach because the government must also be liable for environmental damage, and the civil liability of the government steers equally from lawful public acts if they generate special and unfair damages to certain individuals.[71]

8.3.6 Environmental Studies and Environmental Impact Assessment (EIA)

Decisions that affect environmental balance and generate environmental and intergenerational conflicts must be informed by the best available knowledge about costs and benefits, externalities, and prospective impacts. Science cannot be neglected by public and private decision-makers. Environmental studies, including the Environmental Impact Assessment (EIA), are specifically designed to assess the effects of a given activity on the environment. These studies help guide decision-making regarding activities and facilities that either actually or potentially degrade ecosystems.

Environmental Impact Assessment is a sort of environmental study. The requirement to carry out such studies prior to the establishment of an enterprise or facility has been underscored in various international conventions. For instance, Principle 17 of the 1992 Rio Declaration stipulates that: "Environmental impact assessment, as a national instrument, shall be undertaken for proposed activities that are likely to have a significant adverse impact on the environment and are subject to the decision of a competent national authority." In this context, we can also refer to the "1991 Espoo Convention on Environmental Impact Assessment in a Transboundary

[70] In this regard: Dawalibi (2000).

[71] On this subject, see: Farias (2024), pp. 267–270; Milaré (2018), pp. 1111–1113; Moreira and Wedy (2019), pp. 157–158.

Context," the "Convention on Biological Diversity," and the "1992 United Nations Framework Convention on Climate Change," inter alia.

Inspired by the National Environmental Policy Act (NEPA) of 1969, from North American law, the Environmental Impact Assessment was introduced in the Brazilian legal system under Art. 10, § 3 of Law n. 6.803/80, for industrial zoning in critical areas of pollution. Later, the National Environmental Policy Act (Law n. 6.938/81), in Art. 9, III, defined the Environmental Impact Assessment (EIA) as an instrument of the National Environmental Policy. Later, it was through the provision of Art. 225, § 1, IV of the 1988 Constitution that the EIA became constitutionally mandatory.

Decree n. 99.274/90, which regulates Law n. 6938/81, connected the EIA to the licensing systems, granting the National Environmental Council (CONAMA) the authority to "establish the basic criteria, according to which environmental impact studies will be required for licensing purposes." Article 1, III of CONAMA Resolution n. 237/97 defines environmental studies as those addressing the environmental aspects related to the location, installation, operation, and expansion of an activity or facility. These studies provide essential support for the analysis of the required license and may take the form of environmental reports, environmental control plans, and projects, preliminary environmental reports, environmental diagnoses, management plans, recovery plans for degraded areas, and preliminary risk analyses.

Article 3, sole paragraph of the aforementioned Resolution, establishes that the environmental agency, verifying that the activity or facility does not potentially cause significant environmental degradation, will define the appropriate environmental studies for the respective licensing procedure. Hence, if the project or activity to be licensed does not result in significant environmental degradation, alternative studies, distinct from the EIA, will be conducted. These studies will be tailored to the specific aspects of the anticipated social and environmental impacts of the project.

The Environmental Impact Assessment (EIA), along with the corresponding Environmental Impact Report (RIMA), is a type of environmental study specifically designed for the licensing of facilities and activities that have the potential to cause *significant environmental impact*. In other words, only in those cases in which the project to be licensed has the potential to generate extraordinary, considerable, or above-normal environmental impacts, should the EIA be prepared. This study represents a "deep diagnosis of the project to be licensed by the environmental agency, confronting it with the probable modifications of the several socioeconomic and biophysical characteristics of the environment."[72]

The Environmental Impact Assessment became a constitutional requirement under the 1988 Constitution, which, in Art. 225, § 1, IV, mandates that the government requires a prior environmental impact study—made publicly available—before the installation of projects and activities that may potentially cause *significant environmental degradation*. Consequently, the absence of the EIA/RIMA in the licensing procedure, when it is verified that the impacts to be caused may be

[72]Cappelli et al. (2013), pp. 118–119.

significant, shall result in the cancellation of the license granted. For this reason, e.g., the Supreme Court has decided on the unconstitutionality of a state rule that waives the preparation of an EIA for a specific category of economic activity, without considering the extent of environmental impacts in each case.[73]

Considering that an Environmental Impact Study can only be required in cases of significant environmental degradation, and given its high cost and complexity, it should be employed judiciously and selectively, preferably for projects of greater environmental impact.[74]

It is worth pointing out that the Environmental Impact Assessment is necessarily performed in the context of environmental licensing. In other words, one can conduct an environmental licensing without an Environmental Impact Assessment, when the impacts are not deemed significant. However, there is no EIA outside the licensing procedure. As the Superior Court of Justice has ruled, the EIA/RIMA is not an end in itself and does not represent the final objective. Instead, it serves as merely one of the initial stages in the process leading to the implementation and operation of the facility.[75]

8.3.6.1 Instances Requiring Environmental Impact Assessment

As we have emphasized, the Environmental Impact Assessment is a type of environmental study especially intended for licensing activities and facilities that may cause more serious and significant impacts. The Brazilian Constitution, in Art. 225, § 1, IV, prescribes that it is up to the public authorities to demand, in the manner prescribed by law, for the installation of works and activities that may potentially cause *significant degradation* of the environment, a prior Environmental Impact Study, which shall be made public. However, the Constitution does not define what is meant by *significant environmental impact*, which can create uncertainty regarding the cases in which this study is required. Environmental legislation tried to reduce this uncertainty, but debates remain.

Article 2 of CONAMA Resolution n. 01/1986 provides a list of activities considered to have significant environmental impact. It is widely accepted in environmental law literature that this list is illustrative, not exhaustive. Consequently, for other activities not explicitly mentioned in this Resolution, an Environmental Impact Assessment may still be required if significant impacts are identified in the specific case.[76] Furthermore, the prevailing view is that this regulation establishes a mandatory minimum, binding to the public administration, which can increase, but not reduce, the number of situations in which the Environmental Impact Assessment

---

[73] Brazilian Supreme Federal Court, ADI 1086, Rapporteur Min. Ilmar Galvão, j. 10 Aug. 2001; ADI 5312, Rapporteur Min. Alexandre de Moraes, j. 25 Oct. 2018.

[74] Milaré (2018), pp. 992–993.

[75] Superior Court of Justice, REsp 1072463/SP, Rapporteur Min. Castro Meira, j. 15 Aug. 2013.

[76] In this regard: Milaré (2018), pp. 993–994; Cappelli (1992), p. 54; Sirvinskas (2017), p. 228.

must be prepared. Thus, the activities found on that list are, for most of the doctrine, covered by an absolute presumption of significant environmental impact.[77]

Regarding situations not covered by CONAMA Resolution n. 01/86 or other regulations, the Brazilian Constitution (Art. 225, § 1, IV) uses an open-ended concept: "works or activities that may potentially cause significant environmental degradation." The determination of such cases depends on the specific circumstances of each individual situation. As Benjamin emphasizes, in Brazil, there are two ways of assessing the significance of an impact. One, *ope legis* (by means of the law), sets an absolute presumption of the need to prepare the Environmental Impact Assessment. This is the case of the hypotheses of Art. 2 of CONAMA Resolution n. 01/86. This stands for a total binding of the public agent. The other way, by not having an explicit legal provision, leaves much of the evaluation to the public agent's discretion.[78]

As previously discussed in the study of activities subject to licensing, there are debates regarding the presence of discretionary power when it comes to open-ended and imprecise legal concepts. In any event, even when a statute or regulation uses vague and imprecise legal terms, it is possible to challenge the application of these concepts in individual cases in court. Hence, if it is demonstrated that the installation and operation of a facility have the potential to cause significant negative environmental impacts, any lack or deficiency in the Environmental Impact Assessment may be challenged in court, possibly leading to the invalidation of the issued license.[79]

8.3.6.2 Publicity of the EIA and the Environmental Impact Report

Article 225, § 1, IV of the Constitution prescribes that the Environmental Impact Assessment (EIA) shall be made public. In order to ensure the transparency of studies and decision-making involving environmental impacts, as well as to provide for public participation, the Environmental Impact Assessment must be followed by the Environmental Impact Report (*Relatório de Impacto Ambiental* or RIMA), which represents an accessible summary of the EIA. The EIA itself is lengthier and more complex than the Report and includes it.[80] As Benjamin explained, the EIA represents the entirety of the study: it is complex, detailed, and often contains language, data, and presentations that are incomprehensible to the layperson. The RIMA, on the other hand, is the most accessible and understandable part of the

[77] In the sense that Art. 2 of CONAMA Resolution n. 01/86 establishes an absolute presumption of environmental significance: Machado (2024), pp. 254–255; Cappelli (1992); Benjamin (1992), p. 51.

[78] Benjamin (1992).

[79] Superior Court of Justice has ruled along these lines: REsp 1330841/SP, Rapporteur Min. Eliana Calmon, j. 06 Aug. 2013; REsp 1279607/PR, Rapporteur Min. Mauro Campbell Marques, j. 06 Dec. 2011.

[80] Machado (2024), pp. 275–279.

study, serving as a communication tool that conveys the EIA's findings to both the administration and the public.[81]

The minimum content of the RIMA is set forth in Art. 9 of CONAMA Resolution n. 01/86, which will reflect the conclusions of the Environmental Impact Assessment. The sole paragraph of Art. 9 of the mentioned Resolution establishes that the RIMA must be presented in an objective and appropriate manner for its understanding. The information must be translated into accessible language, and illustrated by maps, charts, graphs, and other visual communication techniques so that the advantages and disadvantages of the project can be understood, as well as all the environmental consequences of its implementation.

Despite the publicity requirement, industrial, commercial, financial, and other legal secrecy must be respected (CONAMA Resolution n. 01/86, Art. 11; Law n. 10.650/03, Art. 2, § 2).

8.3.6.3 Public Hearings

The holding of public hearings with communities interested in a specific economic activity serves as an important tool for public participation in environmental decision-making. According to Art. 11, § 2 of CONAMA Resolution n. 01/86, the appropriate authority must set a deadline for receiving comments from public authorities and other interested parties. Additionally, when deemed necessary, the agency must hold a public hearing to provide information about the project, its environmental impacts, and to discuss the report. A public hearing is not a decision-making procedure. There is no vote on the project, and its conclusion is not binding to the environmental authority. It is a mechanism to strengthen access to environmental information and popular participation, which provides control over the government's decisions. It is of an advisory nature, not deliberative. Nevertheless, the information gathered must be considered by the licensing agency.[82]

The public hearing is currently governed by CONAMA Resolution n. 09/87. Article 1 of this Resolution states that the purpose of the public hearing is to expose to interested parties the content of the project under analysis and its Environmental Impact Report (RIMA), settling doubts and collecting from those present criticisms and suggestions.

However, not all licensing applications require a public hearing. Currently, pursuant to Art. 2 of CONAMA Resolution n. 09/87, whenever the environmental agency deems necessary, or when it is requested by an association, by the Public Prosecution Office, or by 50 or more citizens, the designated authority shall compellingly hold a public hearing. In this case, the request thus formulated binds the

[81] Benjamin (1992).

[82] In this regard, Antunes stresses that the public hearing does not have a decision-making nature. It is an activity of a consultative nature. It is, however, an official act and that, in this condition, must have its results taken into consideration by the licensing agency (Antunes (2023), pp. 155–156).

environmental agency. If the public hearing is not held in this case, any license granted should be invalidated (CONAMA Regulation n. 09/87, Art. 2, § 2).

It is worth observing that the environmental licensing bill (PL 2159/21, not yet approved), mentioned above, proposes to prescribe that a public hearing is required in all environmental licensing procedures for activities or facilities subject to an Environmental Impact Assessment. This will, of course, broaden the cases in which public hearings are mandatory.

8.4 Land Use Regulation and Environmental Zoning

Environmental zoning plays a significant role in ensuring the sustainable use of land and the protection of ecosystems. In addition to helping manage the risk of environmental damage and catastrophes, it limits the rights of individuals in order to protect the environment. Environmental zoning is the result of a deliberation made by the government, to define coexistence among the different interests in the use of geographic spaces, recognizing and solving land use conflicts.[83]

Zoning is an important tool of National Environmental Policy, mentioned under Art. 9, II of Law n. 6938/81. It can be performed at the national, regional, and municipal levels. Zoning represents one of the most significant aspects of administrative police power, exercised to ensure public health, tranquility, peace, and overall well-being. The primary function of zoning is to subsidize planning processes, land use and occupation regulation, as well as the use of environmental resources, especially when creating guidelines for future occupations and licensing.

The Federal Government holds the authority to formulate and implement national and regional plans for territorial organization and for economic and social development (Constitution, Art. 21, IX). However, states and municipalities can and should also regulate zoning matters concerning their own territories. This is not only because it results from the shared responsibility for environmental protection (Constitution, Art. 23, III, IV, VI, and VII), but also because municipalities are empowered to promote appropriate territorial planning through the regulation and control of land use, division, and urban land occupation, wherever applicable (Art. 30, VIII).

In any case, Supplementary Law n. 140/11 assigns to the three levels of the Federation the authority to establish zoning and regulate land use: the Federal Government for national and regional environmental zoning (Art. 7, IX); the states for state-level zoning, in accordance with national and regional zoning (Art. 8, IX); and the municipalities to prepare the Master Plan, observing environmental zoning (Art. 9, IX). One should bear in mind that the federal district encompasses both municipal and state competence.

[83] Antunes (2023), pp. 544–547.

Rules about environmental zoning are also laid down in Law n. 6.803/80, which provides the basic guidelines for industrial zoning and stipulates that the states must establish zoning legislation in critically polluted areas, to make industrial activity compatible with environmental protection; Law n. 9985/2000, specifically about nature conservation units and their surroundings; Law n. 8.171/91, Art. 19, III, which deals with agro-ecological zoning; Law n. 7.661/81, about the regulation of land use and activities in the coastal zone; Decree n. 4.297/02, about Ecological-Economic Zoning; and Decree-Law n. 1.413/75, on the pollution control from industrial activities.

Law n. 6803/80 establishes three categories of industrial zoning, with their respective requirements: (a) strictly industrial use zones; (b) predominantly industrial use zones; and (c) diversified use zones (Art. 1, § 1). These categories may be divided into subcategories, observing the peculiarities of the critical areas to which they belong, and the nature of the industries installed there (Art. 1, § 2).

Strictly industrial use zones are preferably reserved for the location of industrial facilities whose solid, liquid, and gaseous residues, noise, vibrations, emanations, and radiation can cause danger to the health, well-being, and safety of the population, even after the application of adequate methods of control and treatment of effluents (Art. 2). Such zones should be located in areas with high effluent assimilation and environmental protection capacity, respecting all legal restrictions on soil use; be located in areas that favor the installation of infrastructure and basic services necessary for its operation and safety; maintain, around it, *green isolation rings* capable of protecting the surrounding areas against possible residual effects and accidents (Art. 2, § 1, I to III). It is forbidden, in the strictly industrial use zones, the installation of any activities not essential to its basic functions, or capable of suffering harmful effects as a result of these functions (Art. 2, § 2).

Zones of predominantly industrial use are preferably reserved for facilities whose processes, when subjected to proper effluent control and treatment methods, do not significantly interfere with other urban activities or disturb the nighttime rest of residents (Art. 3). These zones should be located in areas where conditions favor the adequate installation of basic service infrastructure necessary for their operation and safety; and must have, within them, environmental protection areas that minimize the effects of pollution, in relation to other uses (Art. 3, sole paragraph, I and II).

Finally, the *zones of diversified use* are intended for the placement of industrial facilities whose production processes complement the activities of the surrounding urban or rural environment and are compatible with them. These facilities, regardless of the use of special pollution control methods, must not cause any inconvenience to the health, well-being, or safety of neighboring populations (Art. 4).

Law n. 6.803/80 also establishes other basic rules about zoning, which are, however, detailed in CONAMA Resolutions[84] and under state, federal district, and municipal legislation.

[84] About CONAMA (National Environmental Council), read in Sect. 7.2.1.

The possibility of zoning changes by the government raises controversy regarding whether a vested right exists for facilities already occupying a site ("pre-occupation") and whether there is a right to compensation if relocation or closure of activities is mandated due to non-compliance with the new regulations.[85]

It should be noted that villages often form around industries, gradually developing into neighborhoods with houses, pharmacies, schools, supermarkets, and other amenities, creating a new community surrounding the facilities.[86] The coexistence between a large polluting source and an expanding urban population eventually becomes detrimental to public health, necessitating, at some point, the adaptation of the facility or, in extreme cases, its shutdown or relocation.

The problem of relocation was dealt with in Decree-Law n. 1.413/75, which provided for the control of environmental pollution caused by industrial activities. It mandates that facilities, whether existing or to be established within the national territory, are required to implement all necessary measures to prevent or remedy any inconvenience and damage caused by environmental pollution and contamination (Art. 1). Furthermore, it stipulates that, in critical areas, an urban zoning scheme will be implemented, aiming to provide suitable relocation alternatives for the most severe cases, even for existing situations. Additionally, it seeks to establish reasonable deadlines for the installation of pollution control equipment in general (Art. 4).

Law n. 6803/80 stated that the "existing facilities or groups of facilities that are not confined in the industrial zones defined in accordance with this statute, will be submitted to the installation of special control equipment and, in the most serious cases, to relocation" (Art. 1, § 3).

It is true that there is no vested right to pollute, nor is the application of the *fait accompli* in environmental law viable. Essentially, it is not acceptable to maintain a polluting activity solely due to the passage of time if it violates the law or if its negative impacts become intolerable for the surrounding community and the ecological balance. However, if a company is in compliance with existing regulations and there are no viable options for adaptation using new techniques and equipment, it would be unfair to mandate the closure or relocation of the facility without offering any compensation. The situation raises questions about property rights, fairness in the distribution of social burdens, and proportionality.

On this matter, the Superior Court of Justice has determined that when a building permit is affected by subsequent zoning rules, the aggrieved party is entitled to government compensation.[87] In a similar case, the same Court ruled that an absolute restriction on property rights goes beyond a mere administrative limitation, thereby entitling the private party to compensation.[88]

Finally, it must be emphasized that zoning rules have their limits. Thus, as the Superior Court of Justice decided, municipal law cannot allow buildings contrary to

[85] Moreira and Wedy (2019), p. 183.

[86] Sirvinskas (2017), p. 222.

[87] Superior Court of Justice, REsp 43.806/MG, Rapporteur Min. Ari Pargendler, j. 14 Aug. 1997.

[88] Superior Court of Justice, REsp 930.589/GO, Rapporteur Min. Sérgio Kukina, j. 12 Apr. 2016.

state zoning laws.[89] In another case, the Supreme Court held that state legislation on environmental zoning must not conflict with federal law, nor can it waive the requirement to prepare an Environmental Impact Assessment broadly for a particular category of economic activity.[90]

8.5 Economic Tools for Environmental Regulation

Economic tools are associated with regulatory mechanisms by inducement and incentive measures. The strategy here is not to punish non-compliance with legislation or to employ the *command-and-control* method. Rather, these instruments pertain to economic incentives for certain sustainable and environmentally protective behaviors, or disincentives for unsustainable and environmentally degrading conduct, thus fostering the voluntary cooperation of private individuals, albeit under inducement.

These economic strategies are widely found in environmental legislation, notably in Law n. 6,938/81, which provides, among the instruments of the National Environmental Policy, "economic tools such as forest concessions, environmental easements, environmental insurance, and others" (Art. 9, XIII).

Among these measures, many of which are market-based, one can mention *green financing*, the Environmental Reserve Quota (addressed in the Forest Protection Code, Sect. 12.3), the carbon market (which will be discussed in the topic on climate change, Chap. 13), environmental taxation, and payment for environmental services.

In this chapter, we will examine two of these mechanisms: environmental taxation and payment for environmental services.

8.5.1 Environmental Taxation

Environmental taxation has increasingly been utilized as a tool to encourage or dissuade specific behaviors, fulfilling a purpose that extends beyond mere fiscal objectives (*extra fiscal function* or *função extrafiscal*). In essence, it involves the use of taxation as a tool to encourage sustainable practices and to internalize the environmental costs associated with economic activities. Also known as *Pigouvian Taxation*, this term references British economist Arthur Cecil Pigou (1877–1959), renowned for his work on externalities—adverse effects of economic activities that are not reflected in the private costs borne by producers or consumers—and for developing public policies aimed at correcting economic distortions.

[89] Superior Court of Justice, AR 756/PR, Rapporteur Min. Teori Albino Zavascki, j. 27 Feb. 2008.

[90] Brazilian Supreme Federal Court, ADI 4069, Rapporteur Min. Edson Fachin, j. 08 Sept. 2020.

Brazilian law offers numerous practical applications of environmental taxation, albeit in a still unsystematic manner.

A notable innovation was the addition of the new item VIII to § 1 of Art. 225 of the Brazilian Constitution by Constitutional Amendment n. 123/22, as amended by Constitutional Amendment n. 132/23. This provision establishes the duty of the government to maintain a preferential fiscal regime for biofuels and low-carbon hydrogen, as defined by Supplementary Law, to ensure a tax rate lower than that applied to fossil fuels. This preferential treatment aims to secure a competitive advantage for these cleaner alternatives. The regulation of this constitutional rule is still pending the approval of Supplementary Legislation.[91]

Other measures to implement environmental taxation include: (a) the transfer of a portion of state taxes to municipalities based on environmental criteria;[92] (b) tax exemptions on rural properties for environmentally protected areas;[93] (c) tax exemptions on cash donations received by federal public financial institutions for actions focused on deforestation prevention, monitoring, and control, including programs that remunerate environmental services and promote conservation and sustainable use of Brazilian biomes;[94] (d) tax reductions for vehicles with higher energy efficiency and lower emissions.[95]

The recent tax reform introduced into the constitution (Constitutional Amendment n. 132/23) has notably aligned the National Tax System with environmental preservation goals: (a) § 3 of Art. 145 now establishes "environmental protection" as a principle of the National Tax System; (b) Art. 153, VIII, grants the Union the authority to impose taxes on "the production, extraction, commercialization, or importation of goods and services harmful to health or the environment, as defined by Supplementary Law"; (c) under Art. 155, § 1, V, the state tax on inheritance and donations, applicable to any assets or rights, "will not apply to donations directed, under the Federal's Executive Branch, to socio-environmental projects or those aimed at mitigating climate change, as well as to federal educational institutions"; (d) according to Art. 155, § 6, II, the tax on motor vehicle ownership may have differentiated rates based on environmental impact.

Environmental taxation, therefore, serves as an economic mechanism increasingly used to encourage sustainable and eco-friendly behaviors or, conversely, to discourage activities that may lead to heightened environmental degradation or greenhouse gas emissions.

---

[91] Currently, Bill n. 68/2024 (*Projeto de Lei Complementar n. 68/*2024) is under consideration in the National Congress, aiming to regulate this new constitutional provision.

[92] Constitution, Art. 158, IV; § 2, II.

[93] Law n. 8171/91, Art. 104; Law n. 9.393/96, Art. 10, § 1, II.

[94] Law n. 11.828/08.

[95] Law n. 14.904/24.

8.5.2 Payment for Environmental Services

Recently, ecologists have been striving to place monetary values on the services that entire ecosystems perform for the benefit of humankind. The substantial costs of replacing many of these services if they were to fail are widely acknowledged. However, the market faces significant limitations in accurately valuing ecosystem services, due to factors like lack of knowledge, institutional inadequacies, and, remarkably, the opportunistic behaviors that arise from public goods. Moreover, concern over the degradation of these services is generally low, as society tends to take them for granted—at least until the day they are no longer available.[96]

The provision of Payment for Environmental Services has therefore evolved as an essential economic instrument to support nature conservation through societal collaboration. It emerged as a mechanism to overcome the traditional invisibility of nature in economic decision-making and to bridge the gap between society and the environment.[97]

The provision of environmental services generally refers to financial and non-financial mechanisms that reward those who maintain, restore, or enhance ecosystem services provided by nature. It is a way to economically recognize the contributions that healthy ecosystems bring to society, such as climate regulation, biodiversity conservation, and watershed protection.

The first payment for environmental services programs emerged as a way to assign economic value to the functions and benefits provided by ecosystems. Frequently cited pioneering cases include the conservation of the Catskill-Delaware watershed in New York in 1990, incentive programs for private forest owners in Costa Rica starting in 1997, and successful practices implemented in Mexico, Colombia, Bolivia, the United States, Australia, and other countries.

In Brazil, the national regulatory framework for payment for environmental services was introduced with Law n. 14,119/21, although prior initiatives already existed at the state and municipal levels.

The new legislation incorporates a distinction between environmental services and ecosystem services. The former refers to human actions that benefit ecosystems, defined as "individual or collective activities that support the maintenance, restoration, or enhancement of ecosystem services" (Art. 2, III). In contrast, ecosystem services represent the "significant benefits for society generated by ecosystems, in terms of maintaining, restoring, or improving environmental conditions," encompassing provisioning, supporting, regulating, and cultural services (Art. 2, II).

Payment for environmental services is defined by Law n. 14,119/21 as a "voluntary transaction in which a payer of environmental services transfers financial resources or other forms of remuneration to a provider of these services under agreed conditions, in compliance with relevant legal and regulatory provisions" (Art. 2, IV).

[96] Portions of this text were partially inspired by: Leape et al. (2018), pp. 116–120.

[97] On the invisibility of nature in the fields of science and economics, particularly within classical economic thought and Environmental Economics, see: Jodas (2021), pp. 7–49.

This agreement includes a *payer* and a *provider*, both encompassing a broad range of potential participants. The *environmental service payer*, according to the statute (Art. 2, V), may be a public authority, non-governmental organization, or private agent, whether an individual or legal entity, at the national or international level. Providers of these services, in turn, may include an individual or legal entity, of public or private law, or a family or community group that, meeting the eligibility criteria, maintains, restores, or improves the environmental conditions of ecosystems (Art. 2, VI).

There are multiple modalities of payment for environmental services, whether monetary or not, including direct payment; provision of social improvements to rural and urban communities; compensation linked to certificates of emission reduction from deforestation and degradation; green bonds; lease agreements; Environmental Reserve Quota;[98] and other arrangements established by regulatory acts of the governing body of the National Policy on Payment for Environmental Services (Art. 3).

Law n. 14.119/21 also introduced the Federal Program for Payment for Environmental Services (PFPSA), to implement the *Federal Policy for Payment for Environmental Services* (*Programa Federal de Pagamento por Serviços Ambientais* or PFPSA) with respect to the payment for these services by the Federal Government, in actions for the maintenance, restoration, or improvement of vegetation cover in priority areas for conservation, combating habitat fragmentation, establishing biodiversity corridors, and conserving water resources (Art. 6).

To benefit from the PFPSA, verification and proof of actions for the maintenance, restoration, or improvement of the subject area are required, in accordance with regulatory provisions (Art. 6, § 6).

A point of controversy lies in the eligibility for remuneration for environmental services funded by public resources, as provided in Law n. 14.119/21, for Permanent Preservation Areas, Legal Reserves, and others under administrative restriction (Art. 9, sole paragraph). This provision is criticized because such areas are already subject to use restrictions and obligations to maintain vegetation, meaning that private parties would benefit merely from complying with the law. Nevertheless, given this possibility established by law, it can be regarded as an additional incentive for compliance with the legislation in force.

On the other hand, the use of public funds for payment for environmental services is prohibited for individuals and legal entities in default with respect to a conduct adjustment or commitment agreement signed with the competent authorities and in relation to areas subject to restrictions by environmental agencies (Art. 10).

The legislation also adopted the *propter rem* nature of the obligation assumed by the property owner in a contract for payment for environmental services, thereby binding the obligation of environmental conservation to the ownership of the property (Art. 22).

[98] On *Environmental Reserve Quota* (*Cota de Reserva Ambiental*), consult Sect. 12.3.3.

The principles of the protector-receiver and user-payer were expressly provided by Law n. 14,119/21 (Art. 5, I). It is worth recalling that the protector-receiver principle requires that the public authorities, instead of punishing harmful outcomes or the commission of infractions, promote incentive measures for behaviors desired by the legal framework through promotional regulations or *reward-based* sanctions.

The legislation also expressly recognized that the maintenance, restoration, and improvement of ecosystem services contribute to the quality of life of the population (Art. 5, II). In other words, environmental protection and the maintenance of *standing forests* (*floresta em pé*) hold significant value for the economy and society.

The complementarity of payment for environmental services in relation to command-and-control instruments was expressly emphasized by Law n. 14.119/21 (Art. 5, IV). Indeed, although repressive mechanisms, also known as command-and-control measures, remain indispensable, public authorities recognize the need for private sector contributions to better protect the environment. Thus, economic approaches that encourage voluntary sustainable behaviors operate alongside and in complement to traditional measures of repressing environmental infractions. To address the persistent crisis in the effectiveness of environmental law, the adoption of economic instruments, particularly payment for environmental services, emerges as an essential method within the context of implementing a participatory and integrated environmental governance system.

References

Antunes PB (2023) Direito ambiental, 23rd edn. Atlas, São Paulo

Benjamin AHV (1992) Os princípios do estudo de impacto ambiental como limites da discricionariedade administrativa. Revista Forense 317:25–45

Cappelli S (1992) O estudo de impacto ambiental na realidade brasileira. Revista do Ministério Público do Estado do Rio Grande do Sul 27:45–60

Cappelli S, Marchesan AMM, Steigleder AM (2013) Direito Ambiental, 7th edn. Verbo Jurídico, Porto Alegre

Dawalibi M (2000) Licença ou autorização ambiental? Revista de Direito Ambiental 17:179–187

Di Pietro MSZ (2024) *Direito Administrativo*, 37th edn. Atlas, São Paulo

Dino Neto N (2011) Introdução ao estudo das infrações administrativas ambientais. Revista de Direito Ambiental 62:169–200

Endicott T (2018) Administrative Law, 4th edn. Oxford University Press, Oxford

Enterría EG, Fernández T-R (1990) Curso de Direito Administrativo. Trad. Arnaldo Setti. Revista dos Tribunais, São Paulo

Farias T (2020) *Competência Administrativa Ambiental*: fiscalização, sanções e licenciamento ambiental na Lei Complementar 140/2011. Lumen Juris, Rio de Janeiro

Farias T (2024) Licenciamento ambiental: aspectos teóricos e práticos, 9th edn. Fórum, Belo Horizonte

Freitas MAP de, Freitas VP de (2014). Direito Administrativo e Meio Ambiente, 5th edn. Juruá, Curitiba

Garcia FA, Moreira Neto DF (2013) A principiologia no direito administrativo sancionador. Revista Brasileira de Direito Público 11:9–28

Jodas N (2021) Pagamento por Serviços Ambientais: diretrizes de sustentabilidade para os projetos de PSA no Brasil. Lumen Juris, Rio de Janeiro

Leape JP, Miller AS, Percival RV, Schroeder CH (2018) Environmental law: law, science, and policy. Wolters Kluwer, New York
Machado PAL (2024) Direito ambiental brasileiro, 30th edn. Malheiros, São Paulo
Mello CAB (2010) Discricionariedade e Controle Judicial, 2nd edn. Malheiros, São Paulo
Mello CAB (2016) Curso de Direito Administrativo, 33rd edn. Malheiros, São Paulo
Milaré É (2018) Direito do Ambiente, 11th edn. Revista dos Tribunais, São Paulo
Moreira RMC, Wedy G (2019) Manual de Direito Ambiental: de acordo com a jurisprudência dos Tribunais Superiores. Forum, Belo Horizonte
Oliveira RF (2014) Ato administrativo, 6th edn. Revista dos Tribunais, São Paulo
Osório FM (2005) Direito administrativo sancionador, 2nd edn. Revista dos Tribunais, São Paulo
Pires LMF (2009) Controle Judicial da Discricionariedade Administrativa: dos conceitos jurídicos indeterminados às políticas públicas. Elsevier, Rio de Janeiro
Sirvinskas LP (2017) Manual de Direito Ambiental, 15th edn. Saraiva, São Paulo
Trennenpohl C, Trennenpohl T (2020) Licenciamento Ambiental, 8th edn. Revista dos Tribunais, São Paulo
Vitta GG (2003) A sanção no direito administrativo. Malheiros, São Paulo
Vitta GG (2008) Responsabilidade civil e administrativa por dano ambiental. Malheiros, São Paulo

Chapter 9
Environmental Civil Liability

9.1 Introduction

Before delving into the study of environmental civil liability, it is essential to clarify two key issues related to Brazilian law.

The first is that the topic of environmental civil liability relates specifically to the *responsabilité civile* in civil law systems or tort law in common law systems, which studies criteria for select situations whether the damage caused to others should be repaired (when it must be repaired) and the way it should be repaired (how it should be repaired).[1] Under the Brazilian Civil Code of 2002, liability can be classified as contractual and extracontractual. A tort can be placed in the second category of extracontractual civil liability, "which is more difficult to be determined, given that, on most occasions, obligations are not previously well established, deriving only from a general duty of abstention. Therefore, it becomes more complex to investigate the damaging conduct, the damage's extent and the causal relation between the damage and the conduct."[2] In Brazil, one aspect of non-contractual civil liability, or tort law, pertains to environmental damage and the restoration of environmental harm.

The second issue to be clarified is about environmental law as a protected right within class actions. In civil law legal systems, three expressions are used to define the type of entitlement, they are, *collective*, *diffuse*, and *homogeneous individual* rights. Such terms are "alien terminology for common law systems." Indeed, these terms do not exist under common law within the scope of class actions or mass tort litigation. However, "the concepts that underlay these labels have both analogues and contemporary application in common law systems."[3] Hence, under Brazilian law, it is safeguarded the rights of collective groups to seek collective relief. These

[1] Battesini (2013).

[2] Celli Jr. (2021), pp. 539–552.

[3] Grinover et al. (2011), p. 265.

R. Martins Costa Moreira, G. Wedy, *Brazilian Environmental and Climate Change Law*, https://doi.org/10.1007/978-3-031-87998-2_9

rights include group-held common interests, such as protection against environmental harm, common nuisance claims, consumer protection, discrimination and employment claims, antitrust violations, inter alia.

Environmental harm arises from the dynamics of a risk society and humanity's growing capacity to impact the surrounding environment. The absence of adequate accountability against environmental degradation results in the privatization of profits and socialization of losses, as well as environmental inequality, especially for the most vulnerable populations subject to the negative impacts of polluting activities.

As Stiglitz has pointed out, not charging industries like coal and oil for the costs they impose on the environment is a *hidden subsidy*, "little different from the other gifts the industry receives in favorable tax treatment and acquiring resources at below fair market prices." Such subsidies arise from a "misalignment between social returns (which may in fact be negative, as a result of the lowering of our standard of living in the wake of environmental deterioration) and private rewards (which are often huge)."[4]

This reality has been considered by the international community and several countries seeking to hold responsible the liable party to pay damages at least to compensate for the environmental harm. In this regard, the Rio Declaration on Environment and Development of 1992, in its Principle 13, prescribes that states shall develop national laws regarding liability and compensation for the victims of pollution and every other environmental harm.

Civil liability for environmental harm in Brazil is usually carried out using class actions, as it will be analyzed in due course. In some cases, however, environmental recovery can be achieved through agreements with governmental entities or the Public Prosecutor's Office.

9.2 Legal Framework for Environmental Civil Liability

Civil liability for environmental harm, under the Brazilian legal system, is based on a strict liability regime, regardless of any intent, negligence, or recklessness, pursuant to Art. 14, § 1 of Law n. 6.938/81. The duty to compensate for environmental harm is drawn from the constitutional text. As established under Art. 225, § 2 of the Brazilian Constitution, those who exploit mineral resources shall be required to restore the degraded environment, in accordance with the technical solution stipulated by the proper public agency, as provided by law. Additionally, according to Art. 225, § 3, activities deemed harmful to the environment shall subject violators, whether individuals or legal entities, to criminal charges and administrative sanctions, along with the obligation to repair the damages caused. The threefold criminal, administrative, and civil liability is enshrined under this constitutional provision, and

[4] Stiglitz (2013), pp. 123–124.

all three are independent, although with some reciprocal influences. Liability regardless of fault is also found in other statutes, such as the Biosafety Act (Law n. 11.105/05, Art. 20), the Solid Waste Act (Law n. 12.305/10, Art. 51), and the Forest Protection Code (Law n. 12.651/12, Art. 2, § 1). In the case of nuclear damage, Art. 21, XXXIII, "d" of the 1988 Constitution expressly provides for liability regardless of any sort of intent, as does Art. 4 of Law n. 6.453/77.

The strict liability scheme for environmental harm still generates academic debate between, basically, two theories: the one of *full risk* and the other of *generated risk*. For the *full risk theory*, the party causing environmental harm cannot claim exemption clauses to escape from the obligation to repair the damage caused. All risks connected to an enterprise should be fully internalized by the productive process, and the liable party should repair all losses that are connected to its activity. On the other hand, the *generated risk theory* seeks to identify, among all risk factors, only the one that, by presenting danger, is effectively capable of generating harmful situations, to impose liability.[5]

The full risk theory has been defended as a matter of fairness since the one who profits from an activity must be held accountable for the risk or disadvantage resulting from his economic business.[6] The Superior Court of Justice has adopted the full risk theory, holding that the causation link is represented by the risk that is established by the inherent risk integrated into the unity of the act. Furthermore, the party causing environmental harm is not allowed to invoke the so-called *liability excluders* or *exemptions clauses*.[7] This means, as the Superior Court has stated in recent precedents, that civil liability for environmental harm, in addition to being strict (regardless of fault), is joint and several among all parties who benefited from the activity that caused the damage.[8]

Thus, the polluter is liable even in case of involuntary damage, and no illegality, foreseeability, or bad faith on his part is ever required, as the pure causation approach is sufficient. The premise is that the entrepreneur must bear the consequences of their hazardous activity.

9.3 Civil Liability for Environmental Harm: Requirements

The fundamental prerequisites for civil liability under environmental law are as follows: (a) the existence of an activity hazardous to health and the environment; (b) actual or potential harm or risk of harm; and (c) a causal connection between the

[5] On the subject, see: Ayala and Leite (2020); Steigleder (2017).

[6] Ayala and Leite (2020), p. 207.

[7] Superior Court of Justice, REsp 1374284/MG, Rapporteur Min. Luis Felipe Salomão, j. 27 Aug. 2014.

[8] Superior Court of Justice, AgInt no AREsp 277.167/MG, j. 14 Mar. 2017.

activity and the harmful outcome. Exemption clauses are not admitted as a rule. The following is a brief analysis of these prerequisites.

9.3.1 Health and Environmental Risk Activity

The operation of an activity that may generate risk to health and the environment is sufficient to constitute liability, regardless of the lawfulness of the act committed. The existence of a valid environmental license or the performance of a lawful activity does not exempt the party responsible for environmental degradation from the duty to repair the damage.

Law n. 6.938/81, Art. 3, IV, brings a broad concept of polluter and degrader, and covers both direct and indirect polluters, be they individuals, companies, or even the government.

If more than one party is responsible for the environmental harm, all are jointly and severally liable to repair the loss. In a lawsuit, a class action plaintiff may bring the claim against one, some, or all polluters at his or her discretion. Subsequently, those who initially pay may file a lawsuit against the other responsible parties to share the burden of compensation.[9]

According to established case law[10] and legislation,[11] the obligation to recover the environmental harm is considered to be vested by a *real right* (right *in rem* or *propter rem*) nature. This means that environmental obligations are tied to the property and can be collected from the current and/or previous owners or possessors. Therefore, if environmental degradation is proven on a given piece of land, even if caused by a previous owner, the current owner is jointly and severally liable for the damage resulting from the contamination and is consequently obligated to undertake remedial and compensatory measures.

The broad scope of civil liability for environmental damage was reaffirmed in a Superior Court of Justice ruling, which stated that those jointly and severally liable include "those who commit the act, those who fail to act when they should, those who permit others to act, those who disregard others acting, those who finance those acting, and those who benefit when others act."[12]

A cause-and-effect relationship between the action and the damage is also required to establish liability for environmental remediation.

---

[9] According to the jurisprudence of the Superior Court of Justice: REsp. 18567/SP; AgInt no AREsp 277.167/MG.

[10] The Superior Court of Justice has settled this understanding in several decisions (AgInt no AREsp 1060669/SP; REsp 1644195/SC) and, finally, consolidated it in the Summula (restatement of case law) (*súmula*) n. 323.

[11] The recognition of the environmental obligation as having a *real right* nature is set forth in the Forest Code (Law n. 12.651/12), Art. 2, § 2; Art. 7, § 2; Art. 66, § 2.

[12] Superior Court of Justice, REsp 650.728/SC, j. 23 Oct. 2007.

9.3.2 Cause and Effect Correlation

The causal link between an activity and its harmful outcomes is a central issue, sparking considerable debate within environmental law. Assessing the causal connection allows us to determine who should bear responsibility for the harmful result and to verify the extent of the damage attributed to the liable party.[13] According to the Superior Court of Justice, considering the application of the *full risk theory*, the so-called *exemption clauses of liability* are not admitted as a rule.[14] In other words, once the causal connection is established, any claim for exclusion of liability is precluded.[15]

In a landmark case decided by the Superior Court of Justice, involving the catastrophic dam burst in the cities of Muriaé and Muriaí in Minas Gerais, and the subsequent massive spill of toxic sludge, the court dismissed the attempt to exclude liability by attributing the disaster to heavy rainfall. The Court declared that the liability for environmental harm must be objective (strict liability), governed by the full risk theory. The causal link is the agglutinating factor that made room for the risk to be integrated into the unity of the act. Hence, it is inappropriate for the company responsible for the environmental harm to invoke exemption clauses or "liability exclusions."[16]

The Superior Court has exempted the defendant from liability, however, when the causal link is not proven.[17] In another case, the same Court held that a landowner is not liable for having his property on fire if the act was committed by a third party completely unrelated to the activity of the former, and the landowner could not be blamed for the damage.[18] Thus, if no harm exists or if it is unrelated to the activity, liability cannot be attributed to those engaged in the activity itself.[19]

On the other hand, in "intervening causes" like *force majeure*, the victim's act or a third party's act is not capable of excluding the causal link. Indeed, the Superior Court of Justice has ruled that liability is not excluded in certain cases, such as: (a) when an activity holds a valid environmental license or operates within emission

[13] Steigleder (2017), p. 173.

[14] Superior Court of Justice, AREsp 1.093.640/SP, Rapporteur Min. Francisco Falcão, j. 15 Mai 2018; REsp 1816808/SP, Rapporteur Min. Herman Benjamin, j. 03 Oct. 2019.

[15] Superior Court of Justice, AgRg in AREsp 232.494/PR, Rapporteur Min. Marco Buzzi, j. 20 Oct. 2015.

[16] Superior Court of Justice, REsp 1374284/MG, Rapporteur Min. Luis Felipe Salomão j. 27 Aug. 2014.

[17] In this case, the Superior Court of Justice ruled out the causal link between the environmental damage resulting from the explosion of a cargo ship and the conduct of the companies acquiring the cargo transported by that vessel (REsp 1596081/PR, Rapporteur Min. Ricardo Villas Bôas Cueva, j. 25 Oct. 2017).

[18] REsp 1381211, Rapporteur Min. Marco Buzzi, j. 19 Sept. 2014.

[19] Steigleder (2017), p. 184.

limits, but impacts arise that were not addressed in the licensing act;[20] (b) the illegal construction or occupation within a permanent preservation area, even if the area is already degraded or anthropized;[21] (c) natural forces, such as a landslide triggered by torrential rains, causing a pipeline rupture and subsequent water pollution;[22] and (d) errors in the issuance of an environmental license.[23]

9.3.3 Environmental Risk and Harm

Not every environmental impact results in environmental liability. For liability to arise, the environmental harm must exhibit a certain level of severity, abnormality, and recurrence, which naturally varies depending on the specific circumstances of each case. In legal proceedings, environmental harm is typically substantiated through expert testimony or reports.

It is also important to understand that ecological harm stands apart from personal and private losses. The so-called *pure ecological harm* is related to, e.g., ecosystem imbalance, pollution and contamination of air, soil, and water, biodiversity loss, etc. It is not necessarily linked to human harm. Both ecological and human damages are equally subject to reparation and may stem from the same event.

Individual damage arising from environmental impact—often referred to as *indirect damage* or *private damage*[24]—takes place, for instance, when fishermen suffer losses from oil spills in the sea or rivers; when people and businesses endure property loss, or even loss of life, due to environmental disasters; or when individuals or communities are deprived of water supply due to river contamination, among other situations.

In a recent case adjudicated by the Superior Court of Justice, it was held that, in light of individual damages sustained by artisanal fishermen due to the conduct of hydroelectric exploitation activities with environmental impact, the legal principle of consumer standing by analogy may be recognized, based on the classification of the incident as a consumer accident. This recognition thereby brings the provisions of

[20] Superior Court of Justice, AgInt no AREsp 1.211.974/SP, Rapporteur Min. Francisco Falcão, j. 23 Apr. 2018; AgInt no REsp 1.419.098/MS, Rapporteur Min. Assussete Magalhães, j. 15 Mai 2018.

[21] Superior Court of Justice, REsp 769.753/SC, Rapporteur Min. Herman Benjamin, j. 08 Sept. 2009; REsp 1.457.851/RN, Rapporteur Min. Herman Benjamin, j. 26 Mai 2015.

[22] Superior Court of Justice, REsp 1346430/PR, Rapporteur Min. Luis Felipe Salomão, j. 18 Oct. 2012.

[23] Superior Court of Justice, REsp 1.612.887/PR, Rapporteur Min. Nancy Andrighi, j. 28 Apr. 2020.

[24] The Superior Court of Justice recognizes the right to compensation for individual damages resulting from environmental impact: REsp 1381211/TO, j. 15 Mai 2014; REsp 1.373.788/SP, j. 6 Mai 2014.

the Consumer Protection Code (Law n. 8.078/90) into effect.[25] Article 17 of this statute considers consumers by analogy (also referred to as bystanders) all victims of an event resulting from defects in products or services rendered.

Compensation for both ecological harm and individual loss is governed by the principle of *strict liability* under the *full risk* theory. The distinction, however, lies in the time limits for initiating proceedings. For the reparation of pure ecological harm, the Supreme Court and the Superior Court of Justice have determined that there is no time limitation.[26] In contrast, for individual damages, the statute of limitations from the Consumer Code (five years)[27] or the Civil Code (three years)[28] applies, depending on whether a consumer relationship exists.

The absence of a statute of limitations for claiming compensation for environmental harm is primarily grounded in the principle that the environment constitutes a common heritage of all humanity, warranting its comprehensive protection, particularly for the benefit of future generations. The prevailing view holds that the reparation of environmental damage is an undeniable fundamental right, justifying the absence of a time limit for filing a lawsuit to seek remedies.

It should also be noted that the theory of *fait accompli*[29] (i.e., an accomplished fact presumed irreversible) does not apply to environmental law. Thus, regardless of how long a structure has stood, if it was illegal at the time of its construction, the owner remains obligated to restore the environment and may even be required to demolish the unlawful structure.[30]

Environmental reparation is guided by the principle of full restoration (*restitutio in integrum*), aiming to repair environmental damage as completely as possible. Prevention of any form of environmental harm is certainly preferred, or, if possible, restoration to its previous state (*status quo ante*). However, as the Superior Court has emphasized, restoration *in natura* is not always sufficient to completely reverse or remedy the damage. The environmental restoration must be achieved in the most complete manner possible, in such a way that an order to recover a degraded area does not preclude the duty to indemnify, especially for the damage that remains between its occurrence and the restoration of the affected environment (= interim or intermediate damage), as well as for the collective nonpecuniary losses and for the

[25] Superior Court of Justice, REsp 2.130.171/SE, Rapporteur Min. Nancy Andrighi, j. 17 Sept. 2024.

[26] Brazilian Supreme Federal Court, RE 654833, Rapporteur Min. Alexandre de Moraes, j. 20 Apr. 2020; Superior Court of Justice, REsp 1394025/MS, j. 21 Sept. 2017; AgRg no REsp 1466096/RS, j. 24 Mar. 2015.

[27] According to Art. 27 of the Consumer Code (Law n. 8.078/90): Superior Court of Justice, REsp 1354348/RS. J. 26 Aug. 2014; AgInt no AREsp 697.506/RS, j. 02 Feb. 2017.

[28] Under the provisions of Art. 206, § 3 of the Civil Code.: Superior Court of Justice, REsp 1641167/RS, j. 13 Mar. 2018; REsp 1346489/RS, j. 11 Jun. 2013.

[29] In Portuguese, *fato consumado*.

[30] Summula (restatement of caselaw) of Superior Court of Justice n. 623. On the subject, in a comparative study of Brazilian and North American law, read on: Marchesan (2019).

residual damage (= environmental degradation that remains despite all the efforts of restoration).[31]

For this reason, the same Court has established that, in a public civil suit, it is possible to cumulatively impose obligations to pay, to act, and to refrain from certain actions. In other words, the judge can, for example, order the defendant to recover the degraded environment, to cease a certain harmful activity, and, cumulatively, to pay a sum of money for the damage caused. Indeed, Art. 3 of Law n. 7.347/85 states that a "public civil suit may aim for monetary compensation or seek the specific performance of an obligation, whether to act *or* to refrain from acting." This expression has been considered by the jurisprudence of the Superior Court of Justice as allowing the judge to decide, cumulatively, for injunctive relief and the payment of damages. The conjunction "*or*" is interpreted as an additive particle and not an excluding one.[32] Thus, the Court affirmed that remedies addressing environmental degradation do not, in themselves, exclude the duty to provide monetary compensation.[33]

However, the combined judicial order for monetary compensation and specific performance is not always mandatory. Courts have clarified that, in certain cases, an award for damages may not be issued if it is possible to fully restore the environment and if the harm is not substantial.[34]

9.3.4 Damages for Collective Nonpecuniary Loss

Damage, as the core concept of civil liability, is not a fixed notion but rather an evolving one.[35] The recognition under Brazilian law of compensation for collective non-pecuniary losses marks another step forward in the evolution of civil liability.[36] These include several types of injury to be compensated, arising from the need for broad protection of personal rights and human dignity, including, according to the predominant view, the recovery of nonpecuniary losses in favor of legal entities.[37]

---

[31] Superior Court of Justice, REsp 1.180.078/MG, Rapporteur Min. Herman Benjamin, j. 02 Dec. 2010.

[32] Superior Court of Justice, Summula (restatement of case law) n. 629; REsp 625.249/PR, j. 15 Aug. 2006. More recently: Superior Court of Justice, REsp 1642723/RS, j. 10 Dec. 2019.

[33] Superior Court of Justice, AgInt no REsp 1659913/SP, Rapporteur Min. Herman Benjamin, j. 31 Mai 2021.

[34] Superior Court of Justice, REsp 1114893/MG, Rapporteur Min. Herman Benjamin, j. 16 Mar. 2010; REsp 1382999/SC, Rapporteur Min. Humberto Martins, j. 19 Aug. 2014; AgInt no AREsp 1217162/RS, j. 18 Mai 2020. In this respect: Mazzilli (2016), p. 174; Moreira (2015); Moreira and Wedy (2019), pp. 453–454.

[35] Martins-Costa (2001), pp. 181–182.

[36] Cavalieri Filho (2023), pp. 108–109.

[37] Summula (restatement of caselaw) of Superior Court of Justice n. 52.

The provision for compensation for collective non-pecuniary losses is based, first and foremost, on the understanding that harm is not solely tied to the victim's pain, distress, or suffering.[38] In the realm of public and collective interests, indemnification stems from the need to protect and preserve fundamental assets and values that are inherently non-economic in nature. These values are connected to the community, and when they are violated, remediation does not require any association with individual elements (affliction, consternation, embarrassment, indignation, humiliation, spiritual shaking, etc.).[39] Nonpecuniary collective losses are recoverable in cases of unfair damage to collective values considered essential to a given community.[40]

Despite the concept's imprecise and ambiguous nature, and the challenges in identifying its manifestation in each case, it is clear that the Brazilian legal system has recognized the compensability of collective non-pecuniary losses. Thus, Art. 5, V and X, of the Constitution does not impose any restriction that would suggest only individual non-pecuniary losses are recoverable.[41] Furthermore, legislation has explicitly established the possibility of compensation for collective non-pecuniary losses, as seen in Art. 6, items VI and VII, of the Consumer Code, and Art. 1 of the Public Civil Suit Act. The Superior Court of Justice acknowledges the possibility of compensating collective non-pecuniary environmental losses, particularly in light of the *polluter-pays* and *restitutio in integrum* principles, given that environmental remediation should be as comprehensive as possible.[42]

The main difficulties in this area involve identifying the presence of a collective non-pecuniary loss in a specific case and determining the appropriate compensation amount. In terms of recognizing such loss, one might ask: What criteria can be used to assess the existence of collective nonpecuniary loss? Or even: is there a way to avoid reliance on the arbitrary discretion of the interpreter and the judge?

In cases of individual non-pecuniary loss, the pursuit of reliable criteria for determining its presence, given the lack of objective standards, often results in applying the principle of proportionality. The solution aligns with the logic of reasonableness and the prevailing ethical-legal standards in society, ensuring that only situations deviating from normality are considered compensable. This approach helps prevent the trivialization or commodification of non-pecuniary damage.[43]

A similar standard is applied in the attempt to identify collective nonpecuniary losses. Legal scholars normally use some criteria such as, for instance, analysis of the "pattern of conduct," "seriousness," and "significant injury;"[44] an impactful and

[38] Martins-Costa (2001), p. 194.

[39] Medeiros Neto (2014), p. 155

[40] Bittar Filho (1994).

[41] Dantas et al. (1996).

[42] Superior Court of Justice, REsp 1.180.078/MG, Rapporteur Min. Herman Benjamin, j. 02 Dec. 2010.

[43] Cavalieri Filho (2023), pp. 109–110.

[44] Medeiros Neto (2014), pp. 186–188.

significant occurrence that exceeds the bounds of tolerability;[45] or, still, a decline in the quality of life and health of the population.[46]

The Superior Court of Justice has decided that collective nonpecuniary losses are recoverable when they exceed the "limits of tolerability, and effectively reach collective values."[47] In the Court's view, "it does not require proof of pain, suffering, and psychological damage, which can be assessed in the sphere of the individual but is inapplicable to diffuse and collective interests." For there to be a duty to indemnify, a situation must occur that implies damage to the "moral values sphere" of a given community, that is, the violation of a collective right.[48]

These parameters, while offering guidance to the judge, provide limited assistance in deciding whether to award compensation for collective non-pecuniary losses in a specific case. The concepts employed are highly vague and imprecise, which calls for the use of the principle of proportionality to identify harm to collective moral values that exceeds normal limits. This approach requires comparison with judicial precedents in similar cases to ensure rationality and equality in the application of law and precedent.[49]

The purposes of collective non-pecuniary loss compensation are also a source of considerable debate. Although ongoing disputes under Brazilian law about the possibility of adopting punitive damages or reparation with a punitive scope,[50] the Superior Court of Justice has held that it is inappropriate to attempt to provide civil reparation for environmental harm with an immediate punitive character, since punishment is a function vested to criminal and administrative law.[51]

Therefore, with the punitive function removed, compensation for collective nonpecuniary harm is now seen as serving a preventive role or even as a means of restoring abstract, collective values. This approach includes accounting for environmental externalities and encouraging more efficient economic activities.[52]

The second major challenge for the indemnification of collective nonpecuniary losses is its valuation. The topic is also subject to a great deal of judicial discretion. Parameters cannot be specified in abstract terms; therefore, it is the judge's

[45] Gajardoni (2014), p. 354

[46] Dantas et al. (1996).

[47] Superior Court of Justice, AgRg no AREsp 809.543/RJ, Rapporteur Min. Humberto Martins, j. 8 Mar. 2016; AgRg no REsp 1.513.156/CE, Rapporteur Min. Humberto Martins, j. 18 Aug. 2015; REsp 1681245/PR, Rapporteur Min. Herman Benjamin, j. 15 Aug. 2017.

[48] Superior Court of Justice, REsp 1402475/SE, Rapporteur Min. Herman Benjamin, j. 29 Mai 2017.

[49] Moreira and Wedy (2019), p. 274.

[50] For the view that compensation for collective nonpecuniary loss may serve a dissuasive and punitive function, see: Ayala and Leite (2020), p. 334; Benjamin (1998). Conversely, for the argument that reparation for moral damage does not entail a punitive function, see: Schreiber (2007); Steigleder (2017), pp. 252–255.

[51] Superior Court of Justice, REsp 1354536/SE, Rapporteur Min. Luis Felipe Salomão, j. 26 Mar. 2014.

[52] Benjamin (1998), Steigleder (2017), pp. 252–255.

responsibility to assess, with reference to judicial precedents in similar cases, the intensity of the risk posed, the severity of the impacts, the economic capacity of the responsible party, and the duration of the degradation.[53]

9.3.5 Destination of the Indemnification Amount

As mentioned above, in the same class action it is possible to order the defendant both to compensate for environmental harm and the damage suffered by individuals or legal entities as a result of the environmental impact. In the latter case, of course, the compensation will be awarded to the injured individual or legal entity.

However, in cases of pure environmental harm, any indemnity awarded—whether for pecuniary or nonpecuniary loss—must be allocated to the *Fund for the Recovery of Damaged Assets*, in accordance with Art. 13 of Law n. 7347/85. This fund is managed by the *Federal Managing Council of the Fund for the Defense of Diffuse Rights*, created by Law n. 9008/95, and regulated by the Decree n. 1.306/94.

The resources collected from the fund will be applied to the recovery of assets, in the promotion of educational and scientific events, and in the publication of informative material specifically related to the nature of the violation or the damage caused, as well as in the administrative modernization of the public agencies responsible for the implementation of environmental public policies (Law n. 9.008/95, Art. 1, § 3).

9.3.6 State Liability for Environmental Damage

Article 3, IV of Law n. 6.938/81 envisages the government as a potential polluter. Furthermore, Art. 225, *caput* and § 1 of the Brazilian Constitution established the duty of the government to protect the environment. Accordingly, this legal framework allows the government to be held liable for environmental harm it may cause, by action or inaction, directly or indirectly. Indeed, according to longstanding jurisprudence, whether the polluter is a private individual or a governmental entity, environmental compensation is governed by the principles of *strict liability* and *full risk*, with the doctrine of joint liability also applying.[54]

The government, even when failing in its duty of oversight and inspection, holds both individual and joint liability alongside the private party that directly caused the

[53] Steigleder (2017), pp. 252–255.

[54] Superior Court of Justice, REsp 1.454.281/MG, Rapporteur Min. Herman Benjamin, j. 16 Aug. 2016.

harm.[55] This happens particularly in the case of failure to comply with the inspection duty by the environmental public agency.[56] The Superior Court of Justice has also stated that, despite the unfeasibility of ubiquitous surveillance, it is necessary to hold the government liable in certain situations, notably when the duty of administrative demolition of an illegal building or other typical self-executing acts that are inherent to the inspection and monitoring prerogative are not timely performed.[57]

However, according to the Superior Court jurisprudence, while the public administration is jointly and fully liable for any environmental damage resulting from its failure to monitor and inspect, this liability is considered to be one of "subsidiary execution." This means that, after both the government and the private party are sentenced to repair the environmental damage, the plaintiff must first seek to enforce the court's order against the private party. Enforcement of the decision against the government is permitted only if the private party is unable to pay the compensation. Thus, there is a priority order in executing the judgment: the private party is charged first, and if it is unable to comply, the government is then held accountable for its failure to act.[58] The reasoning behind this approach is straightforward: it prevents the government from bearing the cost of damages caused by private parties and ensures that the polluter directly internalizes the negative externalities they have created.[59]

References

Ayala PA, Leite JRM (2020) Dano Ambiental, 8th edn. Forense, Rio de Janeiro

Battesini E (2013). Comparative tort law and economics: strict liability in brazilian legal practice. SSRN Electronic Journal. https://doi.org/10.2139/ssrn.2263506.

Benjamin AHV (1998) Responsabilidade civil pelo dano ambiental. Revista de Direito Ambiental 9:5–52

Bittar Filho CA (1994) O dano moral coletivo no atual contexto jurídico brasileiro. Revista de Direito do Consumidor 12:44–62

Cavalieri Filho S (2023) Programa de responsabilidade civil, 16th edn. Atlas, São Paulo

Celli U Jr (2021) Tort law in Brazil. In: Bussani M, Sebok AJ (eds) Comparative tort law: global perspectives. Edward Elgar, Massachusetts, pp 539–552

Dantas MB, Fernandes DCV, Morato Leite JR (1996) O dano moral ambiental e sua reparação. Revista de Direito Ambiental 4:61–67

Gajardoni FdaF (2014) Comentários à Lei de Improbidade Administrativa: Lei 8.429, de 02 de junho de 1992. 3rd edn. Revista dos Tribunais, São Paulo

[55] Superior Court of Justice, REsp 1236863/ES, Rapporteur Min. Herman Benjamin, j. 12 Apr. 2011.

[56] Superior Court of Justice, AgRg no REsp 1417023/PR, Rapporteur Min. Humberto Martins, j. 18 Aug. 2015.

[57] Superior Court of Justice, REsp 1376199/SP, Rapporteur Min. Herman Benjamin, j. 19 Aug. 2014.

[58] Superior Court of Justice, REsp 1376199/SP, Rapporteur Min. Herman Benjamin, j. 19/08/2014.

[59] Superior Court of Justice, REsp 1071741/SP, Rapporteur Min. Herman Benjamin, j. 24/03/2009.

Grinover AP, Mullenix L, Watanabe K (2011) Os processos coletivos nos países de civil law e common law: uma análise de direito comparado, 2nd edn. Revista dos Tribunais, São Paulo

Martins-Costa J (2001) Os danos à pessoa no direito brasileiro e a natureza da sua reparação. Revista da Faculdade de Direito da UFRGS 19:181–207

Mazzilli HN (2016) A defesa dos interesses difusos em juízo: meio ambiente, consumidor, patrimônio cultural, patrimônio público e outros interesses, 29th edn. Saraiva, São Paulo

Medeiros Neto XT (2014) Dano moral coletivo, 4th edn. LTr, São Paulo

Moreira RMC (2015) O princípio da adstrição na jurisdição civil ambiental. Revista de Direito Ambiental 77:169–195

Moreira RMC, Wedy G (2019) Manual de Direito Ambiental: de acordo com a jurisprudência dos Tribunais Superiores. Forum, Belo Horizonte

Schreiber A (2007) Novos paradigmas da responsabilidade civil. Da erosão dos filtros da reparação à diluição dos danos. Renovar, Rio de Janeiro.

Steigleder AM (2017) Responsabilidade civil por danos ambientais: as dimensões do dano ambiental no direito brasileiro. Livraria do Advogado, Porto Alegre

Stiglitz J (2013) The price of inequality: how today's divided society endangers our future. Norton, New York

Chapter 10
Environmental Criminal Liability

Environmental criminal liability is also enshrined under Art. 225, § 3 of the Brazilian Constitution. This constitutional provision is currently developed by Law n. 9.605/98 (Environmental Crimes Act), applied jointly with the Criminal Code and the Criminal Procedural Code. Law n. 9.605/98 establishes a comprehensive framework for criminal liability related to actions deemed harmful to the environment.

Environmental criminal legislation frequently employs *blank criminal laws* (*leis penais em branco*)—criminal provisions that require additional specification through another statute or administrative regulation to be fully understood. This approach arises from the technical complexity inherent in environmental issues. For instance, the crime of "killing, chasing, hunting, catching, using species of wild fauna, native or in migratory route, without the proper permission, license or authorization from the competent authority, or in disagreement with the obtained," and committed "against rare species or endangered species" (Law. n. 9.605/98, Art. 29), requires, for its full understanding, knowledge of the authorization or permit, as well as of the list of rare or endangered species, prepared by a responsible administrative agency. Or even, to verify if the crime of "destroying or damaging a forest considered of permanent preservation" (Art. 38) has been committed, it is necessary to consult the legislation that defines these protected locations and the protection rules.

Criminal prosecution against environmental crimes is *public* and *unconditioned* (Art. 26 of Law n. 9.605/98), i.e., the Public Prosecutor must initiate a criminal lawsuit regardless of any other condition such as the victim's demanding a charge or by any public or private agency. In addition, a criminal lawsuit is not barred by any civil settlement for damages and does not need to await the outcome of administrative proceedings.

The issue of individual criminal liability warrants closer examination. According to Art. 2 of Law n. 9.605/98, anyone who participates in committing crimes addressed by this statute is subject to penalties in proportion to their culpability. This includes directors, administrators, board members, technical staff, auditors, managers, representatives, or agents of a legal entity, as well as those who, despite

R. Martins Costa Moreira, G. Wedy, *Brazilian Environmental and Climate Change Law*, https://doi.org/10.1007/978-3-031-87998-2_10

knowing of others' criminal actions, fail to prevent them when they have the power to do so.

However, this rule does not establish strict (objective) criminal liability. Instead, it establishes criminal liability by omission, arising from the duty of care vested in the individual as a guarantor. In Brazil, for criminal punishment to be imposed, there must be proof of fault (negligence or recklessness) or intent; a person cannot be criminally punished solely for the consequences of their actions. In other words, there must be a guilty mind steering the criminal conduct; that is, it does not exist strict liability under criminal law, in Brazil.

Article 225, § 3 of the Constitution and Law n. 9.605/98 introduced a significant development to Brazilian law by allowing legal entities to be held criminally liable for environmental crimes. Pursuant to Art. 3 of Law n. 9605/98, legal entities are held criminally liable when a violation is committed by the decision of their legal or contractual representative, or a collective body, in the interest or for the benefit of the entity. Additionally, as stated in the sole paragraph of Art. 3, the liability of legal entities does not exclude the liability of individuals involved in the same act.

This legal provision has sparked controversy, with many scholars rejecting the notion of convicting a company for criminal acts. Jurisprudence, however, has generally upheld the feasibility of holding legal entities criminally liable for environmental crimes, as supported by constitutional and statutory provisions.

In an emblematic ruling, the Federal Regional Court of the 4th Region[1] affirmed that, from a criminal perspective, a legal entity can indeed commit an offense. This view is accepted within the legal framework, as legal entities operate within society, and actions carried out in their name and for their benefit can constitute criminal behavior.[2]

The Superior Court of Justice, in the same way, has admitted criminal liability of companies or legal entities in cases of an environmental offense.[3] However, it is important to note that legal entities criminal liability, limited as it is to environmental crimes, does not reach other offenses involving corporations. In fact, there are no constitutional or legal provisions for criminal liability of legal entities in cases other than those found in environmental law.

There has been debate over whether the criminal liability of legal entities requires concurrent criminal prosecution of the individuals responsible for decisions or actions deemed to be environmental crimes. Some scholars and courts have defended the so-called thesis of *double imputation*, according to which the criminal

[1] The Federal Regional Court is the court of appeal in Brazilian Federal Justice. There are currently six federal courts. The one for the 4th Region embraces the three southern states of the country.

[2] Federal Regional Court of 4th R., MS 2002.04.01.013843-0, Rapporteur Fábio Bittencourt da Rosa, j. 10 Dec. 2002.

[3] Superior Court of Justice, REsp 564.960/SC, Rapporteur Min. Gilson Dipp, j. 02 Jun. 2005; REsp 1329837/MT, Rapporteur Min. Sebastião Reis Júnior, j. 08 Sept. 2015.

liability of the legal entity requires the simultaneous liability of the individual who performs the criminal conduct on behalf of the company.[4]

Nevertheless, the Supreme Court has ruled that the Brazilian Constitution does not provide for the *double imputation theory*, meaning that a legal entity's criminal liability for environmental crimes does not require concurrent criminal prosecution of the individual allegedly responsible for the criminal conduct within the organization. The Court stated that today's complex corporate organizations are characterized by decentralization and distribution of tasks and responsibilities, making it more difficult to assign the violation to a particular person.[5] After this decision, the Superior Court of Justice overruled its previous jurisprudence and followed the same path.[6]

The government's liability for environmental crimes is still ripe for controversy. Many commentators sustain the impossibility of criminal liability of the state and public entities, for the following reasons: (a) the sanctions imposed on legal entities are incompatible with the purposes of the state (for example, the suspension or temporary interdiction of activities in light of the principle of continuity of public services); (b) there could be some confusion between the prosecuting body and the person on whom the punishment falls; (c) criminal punishment of the government would represent a "self-flagellation" because it would consist on punishing society itself, which should benefit from the criminal legislation; (d) the requirement of Art. 3 of Law n. 9.605/98 would never be fulfilled, consisting in the commission of the crime in the interest or benefit of the company itself, which contrasts with the purpose of serving the public interest that should guide the behavior of the public administration; (e) if the civil servant performs any act that is not linked with the public purpose, he acts with abuse of power and only the individual will be criminally liable for this act.[7]

Another perspective supports the possibility of criminal punishment against the state, as neither the Constitution nor Law n. 9.605/98 distinguishes between private and public legal entities in matters of criminal liability. Moreover, these scholars argue that this measure encourages greater efficiency within public administration and fulfills a preventive function.[8]

The Superior Court of Justice and the Brazilian Supreme Court have not yet decided on the matter. However, there are some individual decisions from members

[4]This was the previous position of the Superior Court of Justice: RMS 20.601/SP, Rapporteur Min. Felix Fischer, j. 29 Jun. 2006; REsp 564.960/SC, Rapporteur Min. Gilson Dipp, j. 02 Jun. 2005.

[5]Brazilian Supreme Court, RE 548181, Rapporteur Min. Rosa Weber, j. 06 Aug. 2013.

[6]Superior Court of Justice, RMS 39.173/BA, Rapporteur Min. Reynaldo Soares da Fonseca, j. 06 Aug. 2015.

[7]In this regard: Cappelli et al. (2013), pp. 250–252; Freitas (2006), pp. 70–71; Figueiredo and Silva (1998); Krebs (2000).

[8]Cruz (2007); Machado (2024), pp. 764–765.

of theses courts and appellate court rulings admitting such criminal liability of the government.[9]

Penalties for individuals are embedded within the specific "criminal types" that define each criminal behavior. These punishments may include designated prison terms and/or fines. While most environmental crimes fall under Law n. 9.605/98, some are addressed in other scattered legislation.

Yet, the majority of environmental crimes do not result in the incarceration of the defendant due to the relatively lenient penalties prescribed by Brazilian legislation. In these instances, it is possible to apply *decriminalization measures*. These measures have been progressively implemented by Brazilian criminal law to reduce imprisonment for minor offenses. They can be implemented either before or after initiating criminal prosecution and may involve a period of probation. This is the case of the *criminal compromise* (*transação penal)*[10] and the *criminal procedural stay* (*suspensão condicional do processo*), which are regulated under Law n. 9099/95 that deals with *petty crimes* (misdemeanors).[11]

On environmental misdemeanors, Art. 27 of Law n. 9.605/98 prescribes that the proposal for "immediate application of rights restriction punishment or a fine" (which comes from the mentioned *criminal compromise* or *transação penal*), provided for under Art. 76 of Law n. 9.099/95, can only be made if the environmental damage had been previously recovered, except in the case of proven impossibility. As for the *criminal procedural stay*, a kind of probation, Art. 28 of Law n. 9.605/98 requires that, prior to the termination of proceedings, the environmental harm must be repaired, as demonstrated by an appropriate report.

Another measure of pre-trial *decriminalization* was recently brought by Law n. 13.964/19,[12] and refers to the *Agreement Not to Prosecute* (*Acordo de Não Persecução Penal*). This deal applies only to criminal offenses committed without the use of violence or any serious threat and with a minimum punishment provided by law of less than four years imprisonment. This legal provision was an attempt to bring Brazilian legislation closer to the *plea bargaining* of US law to make room for punishment without trial. This agreement takes place between the prosecutor and the defendant, in which the latter confesses to the crime, receives a lighter penalty than the hypothetical final sentence, and, among other conditions, is required to repair the

[9] Individual decision by Min. Rosa Weber from Supreme Court, RE 1307621/SP, j. 04 Feb. 2021, and by Min. Jesuíno Rissato, from the Superior Court of Justice, j. 24 Aug. 21. Within the appellate courts, for instance: Federal Regional Court of 4th Region, ACR 0000574-90.2009.4.04.7200, Rapporteur Salise Monteiro Sanchotene, j. 01 Jul. 2014.

[10] It is a form of plea bargaining, but with some differences, as it does not require a confession from the person under investigation, nor does it carry criminal consequences—except that this measure cannot be applied within five years of the individual's previous use of this benefit.

[11] The measures of *criminal compromise* (*transação penal*) and *criminal procedural stay* (*suspensão condicional do processo*) are ruled under Law n. 9099/95, Arts. 76 and 89, regarding petty crimes (the case of most environmental crimes).

[12] Law n. 13.964/19 amended the Criminal Procedure Code and added Art. 28-A regarding the *agreement not to prosecute*.

harm caused. In the case of an environmental crime, this specifically includes repairing the environmental damage.

If the judicial system proceeds with a criminal trial, the defendant may benefit from replacing a prison sentence with a rights-restriction punishment or other alternative punishment. This option, which was introduced in Art. 44 of the Criminal Code by Law n. 9.714/98, applies to prison sentences of no more than four years for intentional crimes and, and all cases, for non-intentional crimes. In such instances, imprisonment may be replaced by other penalties, such as community service or a fine. These alternative punishments are also governed by Art. 7 of Law 9.605/98 for environmental crimes. Thus, these provisions should be applied first, with the Criminal Code serving in a supplementary role.

Some differences exist under Law n. 9.605/98 on the matter, such as the possibility of applying the substitute punishment only for sentences of less than four years (or any sentence, for non-intentional crimes) (Law n. 9.605/98, Art. 7, I). In addition, Art. 9 of this statute mentions that community service consists of assigning the convicted person free tasks in public parks and gardens and conservation units, and, in the case of damage to private, public, or protected property, its restoration, if possible.

Although the criminal liability of legal entities is recognized, it is clear that the penalties applied to them cannot always be the same as those for individuals, particularly imprisonment. Consequently, penalties must be adapted to suit the specific characteristics of companies. Article 21 of Law n. 9.605/98 establishes that punishments applicable to legal entities—whether separately, cumulatively, or alternatively—include: I - fines; II - restrictions of rights; and III - community service. These penalties are further detailed in Arts. 22 and 23 of Law No. 9.605/98.

References

Cappelli S, Marchesan AMM, Steigleder AM (2013) Direito Ambiental, 7th edn. Verbo Jurídico, Porto Alegre

Cruz GMS da (2007). Responsabilidade penal da pessoa jurídica de direito público nos crimes contra o meio ambiente: uma visão pragmática. Revista Doutrina da 4ª Região. 18: jun. 2007. https://revistadoutrina.trf4.jus.br/artigos/Edicao018/Gysele_Cruz.htm. Accessed 18 Nov 2024

Figueiredo GJP, Silva ST (1998) Responsabilidade penal das pessoas jurídicas de direito público na Lei 9.605/1998. Revista de Direito Ambiental 10:42–59

Freitas VP (2006) Crimes contra a natureza, 8th edn. Revista dos Tribunais, São Paulo

Krebs P (2000) A (ir)responsabilidade penal das pessoas jurídicas de direito público interno. Revista dos Tribunais 772:485–495

Machado PAL (2024). Direito ambiental brasileiro. 30th edn. Malheiros, São Paulo

Chapter 11
Environmental Law in Court

11.1 Introduction

After a thorough discussion of the main issues of Brazilian environmental law, we ought to start studying environmental law enforcement, or how environmental disputes are resolved by the Judiciary. Important note: we will not deal here with criminal lawsuits, studied in the previous chapter. First, let us outline an overview of the Brazilian judicial system.

Since the first Republican Constitution in 1891, due to North American and English influence, Brazil has adopted the *unicity of jurisdiction system*, or the *English system*, which, unlike the French system, submits all conflicts to the Judiciary. In Brazil, there is not, nor there has ever been, a French-style administrative litigation system, so that all disputes, including those involving governmental acts and public law, are decided by the same judicial branch. There was only one failed attempt at creating administrative jurisdiction through a constitutional amendment in 1977, which, however, was never put into practice and was not kept by the 1988 Constitution.

The current constitutional system, therefore, allows broad judicial control of the public administration's actions and inactions, including in environmental matters. Besides the right of access to justice (Art. 5, XXXV), the Constitution enshrines the separation of powers, the independence of the Judiciary, and provides extensive guarantees to judges (Brazilian Constitution, Art. 2; 34, IV; 36, I; 85, II; 92).

The Brazilian Judiciary is divided between two federative levels: federal and state (and federal district) courts. At the state level, there is the *State Court*, which adjudicates all cases, on a residual basis, that do not fall under the jurisdiction of other divisions of the Judiciary. Indeed, the majority of cases are decided by the state courts. At the state level, we still have a specific Military Court to decide criminal cases involving military policemen. It is important to clarify that we are not referring to the Army, as policing activities in the states are divided between the Civil Police—primarily responsible for investigations and inquiries—and the Military

R. Martins Costa Moreira, G. Wedy, *Brazilian Environmental and Climate Change Law*, https://doi.org/10.1007/978-3-031-87998-2_11

Police, which handles visible, preventive policing. In any case, the Military Justice system is of little interest to environmental law.

At the federal level, we have four specialized branches: the Military Justice (with jurisdiction for cases involving members of the Armed Forces); the Electoral Justice (specific for disputes about elections and candidacies for political offices); the Labor Justice (for disputes related to employment relations); and the Federal Justice (which has jurisdiction over both criminal and non-criminal cases involving the participation or interests of the federal government, as well as other national or international interests[1]).

The intention here, once again, is only to provide our readers with an overview of the Brazilian judicial system, without going into its details. The organization of the Brazilian Judiciary has been set forth under Arts. 92 to 125 of the 1988 Constitution.

In Brazil, the same laws on civil and criminal procedure apply in both federal and state courts. Thus, for the prosecution of crimes against the environment, which can occur in state or federal courts, depending on the case, the Criminal Procedure Code, the Environmental Crimes Act (Law n. 9.605/98), and every other piece of legislation can be applied. In every other non-criminal litigation, which we consider as "civil cases" in a broad sense (including, for example, lawsuits on civil law, contracts, torts, administrative law, tax law, non-criminal environmental law, etc.), both federal and state judges are bound by the Civil Procedure Code, the Public Civil Suit Act (Law n. 7.347/85), among other statutes, all equally enacted by the National Congress.

Finally, it is worth mentioning that in Brazil almost every lawsuit is initially processed and judged by a single judge, called *first degree* or *first instance judge*, be it federal or state. Appeals against the decisions of this single judge can be filed in the *High Courts* (*Tribunais de Justiça*) or *Federal Circuit Courts* (*Tribunais Regionais Federais*), which serve as appellate courts, where three high court judges (called *desembargadores*) decide on the appeal. The superior courts have a more limited role, focusing solely on resolving disputes related to federal legislation (handled by the Superior Court of Justice) or constitutional provisions (handled by the Supreme Court).

11.2 Environmental Courts

The critical role of the Judiciary in enforcing environmental law and resolving environmental disputes has been extensively studied worldwide. In a report entitled "Environmental Courts & Tribunals: A Guide for Policy Makers," published by UNEP (United Nations Environment Programme) in 2016, it was stated that "improving the environmental rule of law, access to justice and environmental

[1] The cases that fall under the jurisdiction of the Federal Courts are set forth in Art. 109 of the Constitution.

dispute resolution is essential for achieving the UN's 2030 Agenda for Sustainable Development and the Sustainable Development Goals (SDGs), particularly SDG Goal 16—'to provide access to justice for all and build effective, accountable and inclusive institutions at all levels'. Specialized Environmental Courts and Tribunals (ECTs) are now widely viewed as a successful way to accomplish this important goal."[2] In the updated 2021 version, it was asserted that through "the enforcement of environmental laws and the settling of environmental disputes," Environmental Courts & Tribunals "help countries meet the objectives of the United Nations 2030 Agenda for Sustainable Development and the Paris Agreement on Climate Change, among other international environmental instruments and commitments. They provide access to environmental justice and remedies, strengthen judicial systems to ensure accountability, and spur legal innovation and reforms."[3]

Access to justice has long been regarded as essential for addressing environmental issues, as affirmed by Principle 10 of the 1992 Rio Declaration. This principle was further developed by the 1998 Aarhus Convention and, at the Latin American level, by the Escazú Agreement of 2018. Finally, the first global report on the Environmental Rule of Law, issued by UNEP in 2019, stated that "countries have sought to enhance their courts by improving access to justice to resolve disputes" fairly and transparently. "Because of the technical nature of environmental matters, over 350 environmental courts and tribunals have been established in over 50 countries around the world, including those established at the regional, provincial, or state level."[4]

Brazil has followed a similar path of increasing access to justice to resolve environmental disputes. The 1988 Brazilian Constitution enshrined, under Art. 5, XXXV, that "the law shall not exclude any injury or threat to a right from the consideration of the Judicial Branch." This provision prevents any exclusion from access to justice, guaranteeing the *universality of jurisdiction* and upholding the constitutional right to adequate and effective judicial protection.

In Brazil, judicial procedures have evolved beyond the resolution of disputes over individual rights, advancing to the effective defense of collective interests. The 1967 Constitution, as amended in 1969, had restricted judicial protection to remedies against any injury to individual rights. The Public Civil Suit Law (Law n. 7.347/85), however, marked a significant milestone in expanding judicial protection for community interests, making considerable progress in safeguarding public rights.[5] Subsequently, the 1988 Constitution, under Art. 5, XXXV, ensured the effectiveness of broad access to justice, without the prior limitation to individual rights. Certainly so, it would be useless to recognize the guiding principles of environmental law if they could not be enforced by the justice system; thus, access to justice is an essential premise for effective environmental protection. The situation in Brazil is a privileged

---

[2] United Nations Environment Programme (2016).

[3] United Nations Environment Programme (2022), p. 5.

[4] United Nations Environment Programme (2019).

[5] Freitas and Freitas (2015), pp. 35–41.

one, with broad access to justice allowing the Public Prosecution Service and other entities numerous opportunities to file lawsuits in defense of the environment. Additionally, any citizen may file a similar complaint through a *people's legal suit* (*ação popular*).[6]

To better understand how environmental law is being enforced by our justice system, let us briefly outline the system of class actions and public interest litigation in Brazil.

11.3 Class Actions and Public Interest Litigation in Brazil

The social and economic transformations that took place during the nineteenth and twentieth centuries forced the law to face the new problems arising from the Industrial Revolution, mass consumption, and the progressive human capacity to impact the Earth's resources. The legislation and the judicial procedure needed to incorporate violations of *new collective rights*, such as consumer rights, equality concerns, protection of minorities and the most vulnerable, the environment, etc.

Regarding the protection of these "new rights," there was a lack of a procedural model addressed to the resolution of collective conflicts, which fostered the development of class actions in the United States and other common law countries, and of collective actions in countries with a civil law tradition.

Common law countries, such as the United States, Canada, Australia, and the United Kingdom, have provisions, rules, statutes, or other means to pursue collective actions across a broad range of substantive claims that include labor disputes, environmental claims, securities regulation, consumer actions, etc. The most familiar form of collective action is representative actions pursued through the procedural mechanism of class action.[7]

Class action is linked so strongly with the unique characteristics of the modern era—mass consumption and mass production, large-scale disasters, corporate mismanagement—that it is easy to believe that it is a product of this era. Yet the roots of modern class action stretch back nearly 1,000 years in England past and its development has been ongoing for several centuries.[8] The greatest influence, however, has been wielded by US law. In the United States in the nineteenth century, class actions arose from the contrast between the prevailing individualistic philosophy and the new conflicts that erupted after the Industrial Revolution, until, still in 1842, *representative actions* were codified for the first time in that country under the Federal Equity Rule 48.

The discussion of collective actions in civil law countries (like Brazil) is focused on the historical development of the concept of *diffuse interests* (*interesses difusos*),

[6] Freitas (2005), pp. 35–41.

[7] Grinover et al. (2011), pp. 253–254.

[8] Davis and Eizenga (2011).

which dates back to Italian jurisprudence in early 1974. The concept of diffuse interest had a social and political dimension, recognizing the group nature of some interests and the possibility of group harm. They are scattered and informal interests in the protection of collective necessities. In short, they may be referred to as quality of life.[9] Diffuse interests, then, are community interests, common to a group of people and no one else.

In Brazil, the concept and legal protection of these diffuse rights were established with the enactment of the Consumer Protection Code in 1990 (Law n. 8.078/90). Three types of collective rights (in a broad sense) were established by this law (Art. 81): (a) *diffuse rights*, understood as trans individual rights, of an indivisible nature, held by indefinite persons who are bound by factual circumstances; (b) *collective rights* (in the narrow sense), meaning trans individual rights, of an indivisible nature held by a group, category or class of people linked to each other or the opposing party by a basic legal relationship; (c) *homogeneous individual rights*, which are divisible and entitled to members of specific classes. These homogeneous individual rights may be enforced by a court when filed by individuals, but they can also be dealt with collectively.

It is important to note that this civil law terminology is alien to common law systems. No such terms exist in the common law to describe collective actions. However, although the terms *diffuse rights* or *homogeneous rights* do not exist in common law, the concepts that underlay these labels have both historical analogs and contemporary applications in common law systems.[10]

In this context, it is paramount to clarify that environmental protection is regarded as a diffuse interest, transcending individual rights and serving the broader societal welfare. Thus, it is evident that the environment may be safeguarded through collective legal actions.

11.4 Legal Lawsuits for Environmental Protection

Broad access to justice in environmental matters is enabled through various legal actions or lawsuits, including the public civil action, the people's legal suit, judicial review, and others. For the sake of simplicity, only the most pertinent legal actions within environmental law will be addressed here.

[9] Grinover et al. (2011), pp. 263–264.

[10] Grinover et al. (2011), p. 265.

11.4.1 Public Civil Suit

The public civil suit is the most commonly used type of lawsuit in courts to protect the environment and address environmental disputes. Its popularity stems from its broad scope and the wide range of entities empowered to file it. Introduced in Brazil by Law No. 7.347 in 1985, this suit has since been refined and increasingly employed.

Article 1 of the Law n. 7.347/85 establishes a broad scope for public civil suits, which was expanded through legislative amendments in 1990, 2001, 2011, and 2014. Currently, it deals with the liability for pecuniary and nonpecuniary losses caused to several rights, namely: to the environment; the consumer; assets and rights of artistic, aesthetic, historical, touristic, and landscape value; any other collective or diffuse interest; for violation of the economic order; the urban order; the honor and dignity of racial, ethnic, or religious groups; to public and social heritage.

Once the lawsuit is filed, the legislation gives the judge a wide range of powers to guarantee the effectiveness of the right claimed. The action may have as its objective from harm prevention, including by means of injunctions, to specific performance orders (such as promoting the repair of a degraded area), or abstention (such as refraining from developing a certain highly polluting activity), or to pay damages (condemnation to indemnify for collective pecuniary or nonpecuniary losses). In this regard, Art. 3 of Law n. 7.347/85 allows the public civil suit to seek either monetary compensation or specific performance—requiring a party to take or refrain from taking a particular action. This provision does not preclude the cumulative application of both remedies, as previously mentioned.

The current legislation, moreover, entitles the judge to adopt any measure to enforce the specific performance of the obligation or for the achievement of equivalent practical results. The expansion of judicial powers was initially enabled by the innovations introduced in Art. 84 of the Consumer Protection Code (Law n. 8.078/90) and by amendments to the former Brazilian Civil Procedure Code in 2002. Today, the New Civil Procedure Code (Law n. 13.105/2015) grants judges the authority to provide specific relief or issue injunctions to secure practically equivalent outcomes (Art. 497).

Furthermore, it is no longer necessary for damage to be present; protection can now extend to future and potential damage, or even to prevent mere unlawful acts, whether current or future, regardless of whether any harm has yet occurred. That is to say, the Brazilian procedural system contemplates the possibility of preventing the performance of an activity for simple illegality, regardless of the damages that may arise from unlawful behavior. These are known as an *Inhibitory Injunction* (*Tutela Inibitória*), which aims to prevent a potential legal breach in the future regardless of the likelihood of damage, and the *Injunction to Remove the Legal Breach* (*Tutela de Remoção do Ilícito*), which is intended to eliminate the effects of an unlawful act that has already taken place. These two procedural tools address exclusively unlawful acts, even if harm is not present, to enhance the protection of the right in question.[11] *Inhibitory Injunction* and *Injunction to Remove the Legal Breach*, now expressly provided for in the New Civil Procedure Code of 2015 (Art. 497), are largely suitable for situations involving the discussion of preventive and precautionary measures.

This legal framework grants Brazilian courts a broad scope of action under environmental law, empowering them to pursue various courses of action to ensure the enforcement of rights and the prevention of environmental damage, including actions against the government. Because of this, *judicial review of environmental public policies* has been admitted. As a result, for example, the Brazilian Supreme Court has already found it possible for a court to compel the government to undertake works to prevent river contamination.[12] Similarly, the Superior Court of Justice has imposed on the municipal public administrator the continued provision of garbage collection services.[13] In addition, as studied above, the government can be held liable for any omission in the duty to monitor environmental violations and punish the wrongdoers, which provides room for the court to order the public administration to take concrete actions to prevent or repair environmental damage. More recently, the Supreme Court directed public authorities to implement various measures to promote the protection of the Amazon and Pantanal regions, including the development of prevention and response plans for fires and deforestation.[14]

Article 5 of Law n. 7.347/85 describes the entities with standing to file a public civil suit. They are the following: (I) Public Prosecution Office; (II) Public Legal Defense; (III) Federal, state, Federal District, and municipal government; (IV) public entities or agencies, or public companies; (V) non-governmental organizations and associations that have been constituted for at least one year and include, among its institutional purposes, the protection of the public and social heritage, environment, consumer, economic order, free competition, rights of racial, ethnical or religious groups or the artistic, aesthetic, historical, touristic, and landscape heritage. However, the requirement of *pre-constitution* for non-governmental organizations and associations may be waived by the judge, when there is outstanding social interest evidenced by the dimension or characteristic of the damage, or by the relevance of the legal asset to be protected (Art. 5, § 4).

If not acting as the plaintiff, the Public Prosecutor must serve as the supervisor of legal compliance (Art. 5, § 1). The public entities and other associations entitled to sue may be empowered as co-plaintiffs (§ 2). In case of unfounded withdrawal or abandonment of the lawsuit by the plaintiff association, the Public Prosecution Service or any other legitimate party shall assume this legal position (§ 3). Facultative joint participation will also be allowed among the federal, federal district, and

[11] Marinoni (2004), pp. 268–269.

[12] Brazilian Supreme Court, RE 417408 AgR, Rapporteur Min. Dias Toffoli, j. 20 Mar. 2012.

[13] Superior Court of Justice, REsp 575.998/MG, Rapporteur Min. Luiz Fux, j. 07 Oct. 2004.

[14] Brazilian Supreme Court, ADO 59/DF, Rapporteur Min. Rosa Weber, j. 03 Nov. 2022; ADPF 743/DF, ADPF 746/DF e ADPF 857/MS, Rapporteur Min. André Mendonça, j. 20 Mar. 2024; ADPF 760/DF e ADO 54/DF, Rapporteur Min. Cármen Lúcia, j. 14 Mar. 2024.

the state's public prosecution, in the defense of the interests and rights dealt with by this law (§ 5).

The defendant in the public civil suit corresponds to the figure of the direct or indirect polluter, be it an individual, company, or governmental entity, as stated by Art. 3, IV, of Law n. 6.938/81. The government or public agency usually assumes the position of a defendant when it is demonstrated that there is a deficiency in the duty of inspection or the issuance of an unlawful license, together with the private party that directly caused the environmental harm. The defendant can also be the *potential polluter*, based on the prevention and precautionary principle.

The legislation still allows lifting the corporate veil to reach the assets of the shareholder of a company that is not economically able to repair the environmental damage. For most cases of disregard of legal entity, it is necessary to prove abuse of power, characterized by its misuse of and/or deviation from its purpose (as provided under Art. 50 of the Civil Code). However, in the case of environmental harm, the disregard doctrine applies in any instance of obstruction of compensation for damages caused to the environment (Law n. 9.605/98, Art. 4).[15]

It is also possible to request interim relief to prevent damage or stop an environmentally damaging activity, at the outset of the lawsuit (Law n. 7.347/85, Art. 12). This measure is important and very frequent in environmental lawsuits, especially considering the irreversibility of environmental impacts, in the light of the prevention and precautionary principles.

The proof of environmental harm, past, present, or future, is an extremely complex matter and is quite controversial. Court jurisprudence permits the *reversal of the burden of proof*, requiring the defendant to demonstrate that their activities will not harm the environment or that no environmental damage has been caused. Still, this legal technique does not solve all conflicts involving environmental harm.[16] Many judges and courts admit the evidence gathered by environmental public agencies and bodies, due to their expertise on the subject. Nevertheless, this is not always possible or fair, considering that the public authority may have an interest in the outcome of the dispute.

Sometimes it is necessary to seek expert advice. This can be, in many cases, extremely challenging and costly. As a rule, if the defendant has asked for the expert testimony to be carried out, it is up to him to secure the amount required to pay the expert's costs (Civil Procedure Code, Art. 82). The situation becomes more complicated if the burden of proof is placed on the plaintiff, as in collective suits, the legislation exempts them from covering costs, expert fees, and other expenses (Law n. 7.345/85, Art. 18).

Court jurisprudence has affirmed that the defendant cannot be required to finance lawsuits filed against them, including covering expenses for expert testimony that

[15] Superior Court of Justice, REsp. 279.273/SP; REsp 1339046/SC.

[16] This view is portrayed in the Summula (restatement of case law) (*súmula*) 618 of Superior Court of Justice, in these terms: "the reversion of the burden of proof applies to environmental lawsuits."

they did not request.[17] The amount that the court can pay for expert advice is very low (in the Federal Court, for example, the amount is R$ 372.80, which may be fixed at a maximum of three times in exceptional cases[18]). However, the cost of the environmental expert's witnesses usually far exceeds this figure, and the expert cannot be forced to work for free or without due payment.[19]

Thus, two distinct paths have been taken by jurisprudence. The Superior Court of Justice has ruled that in cases where expert testimony is requested by the plaintiff in a public civil suit or ordered by the judge on their own initiative, the costs should be covered by the government entity to which the Public Prosecutor's Office or the judge (whether state or federal) is attached.[20] However, in a decision by a single judge of the Supreme Court (Min. Ricardo Lewandowski), he stated that if the expert testimony is requested by the Public Prosecutor, the costs must be secured by the Public Prosecution Office itself, using its own resources, as per Art. 91 of the Civil Procedure Code.[21] This remains, however, an isolated decision, without broader precedent in the Supreme Court. Therefore, the prevailing view remains that of the Superior Court of Justice.

When ruling on a case, the judge may order all necessary measures to enforce the recognized right and ensure compliance with the court's decision (Civil Procedure Code, Art. 139, IV). In cases involving *diffuse rights* (such as *environmental rights*), the effects of the final ruling (known as *res judicata* effects) are *erga omnes*, meaning they apply universally across the entire territory covered by the decision, potentially extending to the entire national territory (Consumer Protection Code, Art. 103, I; Law n. 7.347/85, Art. 16). In this regard, the Supreme Court has declared unconstitutional a statutory provision that restrains the effects of the holding issued in a public civil suit to the limits of the territorial jurisdiction of its district.[22]

The parties may also resolve the dispute through an agreement at the beginning, during, or even after the judgment, in this case as a consensual method to facilitate enforcement of the judgment. This agreement can be reached through negotiation, conciliation, mediation, and other consent-based solution approaches, which are

[17] Superior Court of Justice, RMS n. 65.193/SP, Rapporteur Min. Assusete Magalhães, j. 20 Apr. 2021.

[18] *Resolução* (Regulation) n. 575/2019 of the Council of Federal Justice (*Conselho da Justiça Federal*). The Council of Federal Justice shall, pursuant to Art. 105, sole paragraph, II of the Brazilian Constitution, under the terms of the law, exercise administrative and budgetary supervision over the Federal Courts of first and second instances, in the quality of the main body of the system, having powers to correct administrative acts, and whose decisions shall have a binding nature.

[19] As decided by the Superior Court of Justice, the exemption granted to the plaintiff in a public civil suit in relation to the advance payment of expert fees cannot oblige the expert to exercise his office free of charge (Superior Court of Justice, REsp 1253844/SC, j. 13 Mar. 2013).

[20] Superior Court of Justice, AgInt no REsp 1.702.151/SP; RMS 55.476/SP; AgInt no RMS 62.390/SP.

[21] Brazilian Supreme Court, ACO 1560; ARE 1.283.040.

[22] Brazilian Supreme Court, RE 1101973/SP, Rapporteur Min. Alexandre de Moraes, j. 7 Apr. 2021.

encouraged by the new Civil Procedure Code (Art. 3, §3; Art. 135-175; Art. 308, § 3). Furthermore, the Mediation Act (Law n. 13.140/15) allows the use of mediation also to solve disputes concerning individual and public rights that admit transactions, such as the case of environmental disputes (Art. 3). In this instance, however, the agreement should be ratified in court, after hearing the Public Prosecutor (Art. 3, § 2). The limits of this environmental compromise and the powers of the negotiating parties are still subject to debate.[23]

11.4.2 People's Legal Suit

The *people's legal suit* (*ação popular*) was initially brought by Law n. 4.717/65 and later enshrined in the Brazilian Constitution. Under the Constitution, every citizen has standing to file a people's legal suit to nullify acts harmful to the public property, property of entities in which government participation, administrative morality, the environment, and historical and cultural heritage. In bringing this claim, the plaintiff is exempt from judicial costs and the burden of defeat, except in cases of proven bad faith (Art. 5, LXXIII). This lawsuit can only be filed by individuals, and proof of citizenship, required to initiate a claim, is demonstrated by presenting an electoral title (Law n. 4.717/65, Art. 1, § 3).

Thus, it is clear that every citizen has the right to file a people's legal suit in defense of the environment. Yet, due to the costs and challenges associated with this type of lawsuit, people's legal suits on environmental issues are relatively uncommon. It is more usual for individuals to report issues to the Public Prosecution or another public entity, which has the structure and resources to initiate a collective claim, typically through a public civil suit.

11.4.3 Constitutional Suits

The Brazilian Constitution established several types of lawsuits to empower the Supreme Court in ensuring that statutory laws and government actions comply with constitutional rules and principles. These actions, which enable the judicial review of the constitutionality of statutes in Brazil, are as follows: *direct suit of unconstitutionality* (*ação direta de inconstitucionalidade*) to challenge a federal or state law or administrative act; *declaratory suit of the constitutionality* (*ação declaratória de constitucionalidade*) to affirms the constitutionality of a federal law or administrative act.; *suit of unconstitutionality by omission* (*ação direta de inconstitucionalidade por omissão*), which addresses cases where the National Congress or the Executive

[23]For an insightful study on the possibilities and limits of environmental agreement in Brazil, compared to other countries, see: Moreira (2021).

Branch fails to implement a constitutionally mandated provision; the *claim of non-compliance with a fundamental precept* (*arguição de descumprimento de preceito fundamental*), which has a broader scope than other constitutional lawsuits, allowing challenges against administrative acts, state and municipal laws, and legislation enacted before the 1988 Constitution.[24]

We do not have space here to delve deeply into the topic of constitutional control over government actions. However, it is worth noting that numerous proceedings challenge government laws, actions, and inactions on environmental issues, primarily based on Art. 225 of the Constitution and its paragraphs. Recently, for instance, the Supreme Court ruled that the federal government has a constitutional, *supra-legal*,[25] and legal obligation to protect the environment and to combat climate change. The Court held that the Executive Branch is constitutionally required to ensure the annual operation of the *Climate Fund* and to allocate resources for climate change mitigation.[26] In the same vein, the Court ordered the reactivation of the *Amazon Fund*, declaring that it is unconstitutional to take any measure that implies the absence of adequate public policies for the protection of the Amazon Rainforest.[27]

11.5 Judicial Deference and the Scope of Judicial Review

In environmental lawsuits, Brazilian courts have numerous tools to enforce environmental law effectively, whether addressing private companies or government actions and inactions.

The legislation gives the Brazilian judge the power to adapt his orders to the circumstances of the dispute, including in a manner different from that requested by the plaintiff. As previously mentioned, Art. 139, IV of the Civil Procedure Code grants judges the authority to employ all necessary inductive, coercive, injunctive, or substitutionary remedies to ensure compliance with court orders, including claims involving monetary benefits. Article 497 of the Civil Procedure Code provides that, in cases where the subject matter involves an obligation to act or refrain from acting, the judge must grant specific relief or order measures to ensure that relief is achieved through a practically equivalent outcome. Such provisions align with Art. 84 of the Consumer Protection Code (Law n. 8.078/90), which states that in lawsuits seeking to enforce an obligation to act or refrain from acting, the judge may issue an order for specific performance or mandate measures to ensure a practical remedy for the harm caused.

[24] These different types of suits do not exist in common law countries, but it is worth mentioning here.

[25] That is, it is a duty provided for in international treaties.

[26] Brazilian Supreme Court, ADPF 708/DF, Rapporteur Min. Roberto Barroso, j. 01 Jul. 2022.

[27] Brazilian Supreme Court, ADO 59/DF, Rapporteur Min. Rosa Weber, j. 3 Nov. 2022.

Accordingly, for instance, if the aim is to halt a polluting activity, the judge may instead order the installation of a specific technology (such as a filter) to stop the pollution, even if this differs from the plaintiff's request. Another example would be illegal construction in an environmentally protected area; even if the plaintiff only requests the restoration of the degraded area, the judge may order the demolition of the building if necessary to restore the affected ecosystem.[28]

This range of judicial measures is even broader in cases of "structural litigation," where it is acknowledged that the primary threat to constitutional values comes not from individuals but from the operations of large-scale organizations. In such cases, it is understood that unless these organizations are restructured, the threats to constitutional values cannot—and will not—be eliminated.[29] The use of structural processes and structural injunctions in Brazilian civil procedure is still in its early stages of study and development. Nonetheless, many commentators have begun to recognize that structural approaches are valuable for addressing environmental disputes. These cases often require dialogue among involved institutions, the adoption of creative measures by the courts, the possibility of partial agreements and decisions, and the formulation of a forward-looking plan.[30]

Two practical cases illustrate this approach. The first is the so-called *coal public civil suit* in Criciúma, where multiple class actions were filed in the Federal Court. Numerous judicial decisions and agreements were reached with various companies and public agencies to restore areas damaged by coal mining and to implement less environmentally harmful resource extraction methods.[31] Another case involves a public civil suit in the Federal Court of Florianópolis, which mandated the adoption of structural measures for establishing a socio-ecological governance system to manage, protect, control, and monitor current and future impacts affecting the ecological integrity of the Conceição Lagoon (*Lagoa da Conceição*), an area heavily impacted by urbanization.[32]

Another issue that deserves explanation is the scope of judicial review of administrative acts of environmental relevance in Brazil. As reported above, Brazil has adopted the "unique jurisdiction system," so that the Judiciary has the prerogative to review and overturn any governmental action, as long as they are illegal, unconstitutional, or contrary to legal principles. In the field of environmental law, courts have granted little deference to the Executive Branch and administrative bodies and agencies. Thus, for example, the Judiciary can review an administrative decision

[28] Moreira (2015).

[29] Fiss (1984).

[30] Studies regarding the use of the structural techniques in Brazilian legal system can be found in: Arenhart (2013); Arenhart and Jobim (2021); Cambi and Wrubel (2019); Didier and Zaneti (2018); Jobim (2013); Jobim (2015); Vitorelli (2020).

[31] Arenhart (2015).

[32] For a brief explanation of the case in English, see: Brotetto and Cavedon-Capdeville (2021).

that waived the requirement for an Environmental Impact Assessment,[33] suspend permits for economic activities with environmental impacts if mitigating measures are lacking or if impacts reach a certain threshold,[34] compel the government to implement environmental protection measures,[35] to mention a few.

In the context of judicial review of the constitutionality of statutes, however, the Supreme Court has shown considerable deference to the legislature regarding the Forest Protection Code, reasoning that the democratic principle should prevail, even if it means upholding legislation that is less protective of the environment.[36] In a more recent case, the Supreme Court decided that the Executive Branch has the constitutional duty to operate and allocate annually the resources of the Climate Fund, to mitigate climate change, due to the constitutional duty to the protection of the environment (Constitution, Art. 225), the international commitments assumed by Brazil (Constitution, Art. 5, § 2), as well as the constitutional principle of separation of powers.[37]

In a series of rulings concerning the Pantanal and Amazon biomes, the Supreme Court held that public authorities do not have the discretion to refrain from adopting effective measures for the protection, fire prevention, and deforestation control of these vital Brazilian biomes. Accordingly, the Court deemed the institutional dismantling of the Amazon Fund unconstitutional,[38] ordered the federal government to assume meaningful engagement in addressing illegal deforestation of the Amazon Rainforest,[39] mandated the development of a plan for fire prevention and control in the Pantanal and Amazon regions,[40] and directed the National Congress to enact a regulatory law for the special protection of the Pantanal biome.[41] It can be concluded that this more interventionist stance by the Supreme Court in these cases was highly influenced by the drastic and alarming problem of fires and deforestation in the Amazon and Pantanal regions. In other words, the current context impacts the level of deference and judicial intervention in environmental public policies.

It is important to note that when a clear constitutional or legal mandate exists, the Judiciary generally enforces this mandate, leaving no discretion for the Executive or Legislative branches. This is known as a *binding power* or *binding decision.* However, in areas marked by legal ambiguity or policy considerations, judicial

[33] Superior Court of Justice, AgRg no AREsp 476.067/SP, j. 22 Mai 2014; REsp 1279607/PR, j. 06 Dec. 2011; REsp 938.484/MG.

[34] In this regard, the Supreme Court has decided to stay the mining activity for noncompliance by the mining company with preventive, mitigating and compensatory measures provided for in the environmental license, having found evidence of serious damage caused to Indigenous communities (Brazilian Supreme Court, SL 933 ED/PA, j. 31 Mai 2017).

[35] Superior Court of Justice, AREsp 2.024.982-SP, Rapporteur Min. Og Fernandes, j. 14 Jun. 2022.

[36] Brazilian Supreme Court, ADC 42, ADI 4901, 4902, 4903, 4937, j. 28 Feb. 2018.

[37] Brazilian Supreme Court, ADPF 347 MC. ADPF 708/DF, j. 1 Jul. 2022.

[38] Brazilian Supreme Court, ADO 59, j. 3 Nov. 22.

[39] Brazilian Supreme Court, ADPF 760, ADO 54, j. 14 Mar. 24.

[40] Brazilian Supreme Court, ADPF 746, ADPF 857, j. 20 Mar. 24.

[41] Brazilian Supreme Court, ADO 63/MS, j. 6 Jun. 24.

deference varies significantly, making it difficult to establish a precise standard for judicial scrutiny. Generally, in cases of environmental harm and decisions by the Executive Branch or environmental agencies, the courts tend to examine the case and conduct a judicial review. However, in the judicial review of legislative acts, some Supreme Court rulings afford deference to the National Congress, though this approach is not uniformly applied.

References

Arenhart SC (2013) Decisões estruturais no processo civil brasileiro. Revista de Processo 225:389–410

Arenhart SC (2015) Processos estruturais no direito brasileiro: reflexões a partir do caso da ACP do carvão. Revista de Processo Comparado 2:211–229

Arenhart SC, Jobim MF (2021) Processos Estruturais, 3rd edn. Juspodivm, Salvador

Brotetto V, Cavedon-Capdeville F (2021) The rights of nature before the courts: the case of Lagoa da Conceição in Brazil. International center for comparative international law. https://cidce.org/en/the-rights-of-nature-before-the-courts-the-case-of-lagoa-da-conceicao-in-brazil/. Accessed 18 Nov 2024

Cambi E, Wrubel VTS (2019) Litígios complexos e processo estrutural. Revista de Processo 295: 55–84

Davis E, Eizenga MA (2011) A history of class actions: modern lessons from deep roots. Can Class Action Rev 7:5–31

Didier Jr F, Zaneti Jr H (2018) Curso de Direito Processual Civil: Processo Coletivo. V. 4. 12th edn. JusPODIVM, Salvador

Fiss O (1984) Two models of adjudication. In: Goldwin RA, Schambra WA (eds) How does the constitution secure rights? American Enterprise Inst. for Public Policy Research, Washington, DC, pp 36–49

Freitas MAPde, Freitas VPde (2015) A complexidade das ações civis públicas envolvendo meio ambiente e populações vulneráveis. In: Milaré É (coord.) Ação Civil Pública após 30 anos. Revista dos Tribunais, São Paulo

Freitas VP (2005) A Constituição Federal e a efetividade de suas normas, 3rd edn. Revista dos Tribunais, São Paulo

Grinover AP, Mullenix L, Watanabe K (2011) Os processos coletivos nos países de civil law e common law: uma análise de direito comparado, 2nd edn. Revista dos Tribunais, São Paulo

Jobim MF (2013) Medidas Estruturantes: da Suprema Corte Estadunidense ao Supremo Tribunal Federal. Livraria do Advogado, Porto Alegre

Jobim MF (2015) A structural reform no direito brasileiro e a atuação democrática do supremo tribunal federal na sua implementação. Revista de Processo Comparado 2:159–179

Marinoni LG (2004) Técnica Processual e Tutela dos Direitos. Revista dos Tribunais, São Paulo

Moreira RMC (2015) O princípio da adstrição na jurisdição civil ambiental. Revista de Direito Ambiental 77:169–195

Moreira RMC (2021) Acordos ambientais e os limites do inegociável. Thoth, Londrina

United Nations Environment Programme (2016) Environmental Courts & Tribunals: A Guide for Policy Makers. Sept. 2016. https://wedocs.unep.org/bitstream/handle/20.500.11822/10001/environmental-courts-tribunals.pdf?sequence=1. Accessed 18 Nov 2024

United Nations Environment Programme (2019) Environmental Rule of Law: First Global Report. Nairobi, 2019. http://wedocs.unep.org/xmlui/bitstream/handle/20.500.11822/27279/Environmental_rule_of_law.pdf?sequence=1&isAllowed=y. Accessed 18 Nov. 2024

United Nations Environment Programme (2022) Environmental Courts & Tribunals - 2021: A Guide for Policy Makers. file:///C:/Users/Rafael%20Moreira/Dropbox/PC%20(3)/Downloads/ECT2021%20(1).pdf. Accessed 18 Nov 2024

Vitorelli E (2020) Processo Civil Estrutural: Teoria e Prática. JusPodivm, Salvador

Chapter 12
Specially Protected Areas

12.1 Introduction

Protected areas or conservation areas are locations that receive special protection because of their well-known natural, ecological, or cultural values. There are several types of protected areas, which vary according to the level of protection, the purpose of their creation, and the regulation of each country or international organization.

When the Parties to the Convention on Biological Diversity adopted Target 11 in 2010, they acknowledged that effective and equitable systems of protected areas are essential tools to address the biodiversity crisis. By 2020, at least 17% of terrestrial and inland water, and 10% of coastal and marine areas, especially areas of particular importance for biodiversity and ecosystem services, were conserved through an effective and equitably managed, ecologically representative, and well-connected system of protected areas and other effective area-based conservation measures and integrated into wider landscapes and seascapes.[1]

Brazil holds about one-third of the world's remaining primary tropical rainforests, including about 60% of the Amazon rainforest. Terrestrially speaking, it is also the most biodiverse country on Earth, with more than 34,000 described species of plants, 1813 species of birds, 1022 amphibians, 648 mammals, and 814 reptiles.

The Amazon rainforest is the world's largest intact forest. It is home to more than 24 million people in Brazil alone, including hundreds of thousands of Indigenous Peoples belonging to 180 different groups. The region is home to 10% of all plant and animal species known on Earth. In addition to its unparalleled diversity of life, the Amazon plays an essential role in helping to control the planet's climate. The Amazon Basin stores approximately 100 billion metric tons of carbon—that is more than ten times the annual global emissions from fossil fuels. While it covers 2.6 million square miles across nine countries—Brazil, Bolivia, Peru, Colombia, Ecuador, Venezuela, Guyana, Suriname, and French Guiana—about 60% of the

[1] United Nations Environment Programme (2020).

R. Martins Costa Moreira, G. Wedy, *Brazilian Environmental and Climate Change Law*, https://doi.org/10.1007/978-3-031-87998-2_12

Amazon Basin is in Brazil.[2] Therefore, the importance of focusing attention on the regulation of environmentally protected areas in Brazil is indisputable.

Although, in Brazil, protected areas have existed since 1937 (the year the Itatiaia National Park was created), it was only under the impulse of the National Environmental Policy, under Law 6.938/81, that a group of conservation units could be established, although on a case-by-case basis, always managed with few resources and lacking a defined environmental policy action.[3] Thus, the institution of specially protected areas by the federal, state, district, and municipal governments was set forth under Law 6.938/81, Art. 9, IV, as a regulatory tool of the National Environmental Policy.

The 1988 Constitution represented a landmark by prescribing, as a duty of the government, to define territorial spaces and their components to receive special protection. In addition, it stated that any alterations or suppressions of this area are legitimate only by means of statutory law, and any use that may harm the integrity of the attributes that justify their protection is forbidden (Constitution, Art. 225, § 1, III). This constitutional provision gave a major impetus to the issuance of appropriate legislation and the expansion of environmental conservation areas.

It is important to note that the creation of a protected area can occur through either statute enactment or administrative regulation. In contrast, the reduction or elimination of protection levels is permitted only through a statute enacted by the Legislature. As the Supreme Court has decided, it is not permissible to reduce or eliminate a conservation unit by means of a *Provisional Measure* (*Medida Provisória*) issued by the President of the Republic.[4]

In Brazil, by 2020, at least 30% of the Amazon, 17% of each of the other terrestrial biomes, and 10% of the marine and coastal areas, especially areas of particular importance for biodiversity and ecosystem services, are conserved through protected areas provided for under the Nature Conservation Units Act (Law n. 9.985/00) and other categories of officially protected areas such as Permanent Protection Areas, private land preservation areas, and Indigenous lands with native vegetation, ensuring and respecting the demarcation, regularization, and effective and equitable management, to ensure ecological interconnection, integration and representation in broader landscapes and seascapes.[5]

In this book, we will address the two main statutes that deal with protected areas in Brazil: Nature Conservation Units Act (Law n. 9.985/00) and the Forest Protection Code (Law n. 12.651/12).

[2] Greenpeace (2024).

[3] Milaré (2018), p. 1544.

[4] Brazilian Supreme Court, STF, ADI 4717/DF, j. 5 Apr. 18.

[5] United Nations Environment Programme (2020).

12.2 Nature Conservation Units' System

Following the 1988 Constitution, the Nature Conservation Units Act (Law n. 9.985/00) was enacted in 2000. This law listed and defined the different categories of nature conservation units. It is defined by the statute as a territorial space and its environmental resources, including bodies of water, with significant natural characteristics, legally designated by the government for conservation purposes. These areas have defined boundaries, are under special administration, and are subject to specific protection measures (Art. 2, I). All three federative levels (federal, state, and municipal) have the authority to create conservation units.

Law n. 9985/00 established two different categories of conservation units: *fully protected conservation units* and *sustainable use conservation units* (Art. 7).

The *fully protected unit*, as the name suggests, consists of areas requiring higher levels of preservation due to their ecological value and vulnerability. Its primary purpose is to preserve nature, allowing only the *indirect use* of its natural resources, as specified by law (Law n. 9985/00, Art. 7, § 1). The statute defines the expression *full protection* as the "maintenance of ecosystems free from alterations caused by human interference, with only the indirect use of their natural attributes being admitted" (Art. 2, VI). *Indirect use*, in turn, is defined as "that which does not involve consumption, collection, damage, or destruction of natural resources" (Art. 2, IX).

There are five types of fully protected units, each with its own features and regulations: I—Ecological Station; II—Biological Reserve; III—National Park; IV—Natural Monument; V—Wildlife Refuge (Art. 8). Not all of them are constituted as government property. The creation of an Ecological Station, Biological Reserve, and National Park requires the compulsory purchase of private lands, since its areas are under the public domain (Art. 9, § 1; Art. 10, § 1; Art. 11, § 1). In turn, Natural Monuments and Wildlife Refuges can be made up of private areas, as long as it is possible to make the objectives of the unit compatible with the use of the land and natural resources of the site by the owners (Art. 12, § 1; Art. 13, § 1).

The goal of the *sustainable use conservation units* is to make nature conservation compatible with the sustainable use of a portion of its natural resources (Art. 7, § 2). The term *sustainable use* is defined by the law as the "environmental exploitation in a manner that ensures the perpetuity of renewable resources and ecological processes, preserving biodiversity and other ecological attributes, in a socially just and economically viable way" (Art. 2, XI). Therefore, the sustainable use unit allows for some level of human use and occupation, imposing less stringent restrictions than those of a fully protected unit.

There are seven types of sustainable use units: I—Environmental Protection Area; II—Area of Relevant Ecological Interest; III—National Forest; IV—Extractive Reserve; V—Fauna Reserve; VI—Sustainable Development Reserve; and VII—Private Natural Heritage Reserve (Art. 14). Each of these areas has a specific regulation prescribed under Law n. 9.985/00. Of these units, only the National

Forest, Extractive Reserve, and Fauna Reserve are under the public domain, and their creation requires the compulsory purchase of private land.

In the case of conservation units that coexist with private properties within them, there is no right, as a rule, to compensation in favor of the owner facing legal limitations. This is because it is a *general binding rule of territorial planning*, which applies to everyone in the same situation.[6]

A conservation unit can be created by any governmental act, whether through a statute or decree, at the federal, state, district, or municipal level (Art. 22). However, the suppression (total or partial) and the reduction of the protection level relies upon specific legislation.[7] Furthermore, it is not illegal to create more than one type of nature conservation unit from a single administrative procedure.[8] Therefore, a more flexible and facilitated procedure is provided for the creation and expansion of conservation units than for their suppression or reduction.

The creation (and the expansion[9]) of a conservation unit must be preceded by technical studies and public consultation that allow the identification of the most appropriate location, size, and boundaries for the unit (Law n. 9.985/00, Art. 22, § 2). In this consultation process, the public authority is bound to provide adequate and intelligible information to the local population and other interested parties (Art. 22, § 3). However, for the creation of an Ecological Station or Biological Reserve, consultation is not mandatory (Art. 22, § 4), due to the more pronounced public interest that underlies the establishment of these protected areas. The purpose of public consultation is to assist in defining the location, size, and boundaries of the protected area.[10] It is not deliberative and does not determine whether a conservation unit will be created.[11]

To provide more efficient protection, conservation units must have a *buffer zone* (*zona de amortecimento*), which is considered to be their surroundings, where human activities are subject to specific rules and restrictions to minimize the negative impacts on the protected area (Law n. 9.985/00, Art. 2, XVIII). Only in relation to the Environmental Protection Area and the Private Natural Heritage Reserve, the buffer zone is not required (Art. 25).

In addition, one or more conservation units may have *ecological corridors* (*corredores ecológicos*), that is, portions of natural or semi-natural ecosystems, connecting conservation units, which allow the flow of genes and the movement of biota between them, facilitating the dispersal of species and the recolonization of

---

[6] Superior Court of Justice, AgInt no AREsp 1551978/SC, j. 11 Feb. 2020.

[7] Law n. 9.985/00, Art. 22, §§ 5, 6, and 7; Constitution, Art. 225, § 1, III. From the Supreme Court jurisprudence: ADI 4717/DF; MS 25284.

[8] Brazilian Supreme Court, MS 25347, j. 17 Feb. 2010.

[9] As the Brazilian Supreme Court has ruled, the expansion of a conservation unit also requires prior public consultation and technical studies (MS 24665/DF).

[10] Decree n. 4.340/02, Art. 5.

[11] Brazilian Supreme Court, MS 25347, j. 17 Feb. 2010.

degraded areas, as well as the maintenance of populations that require, for their survival, larger areas than those of individual units (Art. 2, XIX).

The Brazilian law also embodies conservation units' areas recognized as Biosphere Reserves by UNESCO (Art. 41). The Man and Biosphere (MAB) Program is an intergovernmental scientific program that aims to establish a scientific ground for the enhancement of the relationship between people and their environment. It combines the natural and social sciences with a view of improving human livelihoods and safeguarding natural and managed ecosystems, thus promoting innovative approaches to economic development that are socially and culturally appropriate and environmentally sustainable. The World Network of Biosphere Reserves currently counts 738 sites in 134 countries all over the world, including 22 transboundary sites.

There are seven regions considered Biosphere Reserves by UNESCO in Brazil: Atlantic Rainforest (*Mata Atlântica*), *Cerrado, Pantanal, Caatinga, Central Amazon, Espinhaço Range*, and *São Paulo City Green Belt*. A Biosphere Reserve is composed of: one or several core areas, designated for integral protection of nature; one or several buffer zones, where only activities that do not result in damage to the core areas are admitted; one or several transition zones, without rigid boundaries, where the occupation process and the management of natural resources are planned and conducted in a participatory manner and on a sustainable basis (Art. 41, § 1). A Biosphere Reserve may be constituted by areas of public or private domain and may be integrated by conservation units already created by the government (Art. 41, §§ 2 and 3). The Biosphere Reserve is managed by a Deliberative Council, made up of representatives from public institutions, civil society organizations, and the local population, as provided for in the regulations and the act of formation of the unit (Art. 41, § 4).

12.3 Forest Protection Code

The protection of forests and vegetation in Brazil goes back to the time of *colonial Brazil*, from 1500 to 1822, of *Imperial-Brazil*, from 1822 to 1889, and the *Old Republic*, from 1889 to 1930. Nevertheless, the vegetation protection in this remote period was scattered and aimed exclusively at safeguarding economic interests related to logging, agriculture, and neighborhood rights. The first forest protection statute, with preservationist intent, was the Decree n. 23.793/34. It was the first true Forest Code in Brazil, still in 1934, which provided for the forest as an asset of common interest, restricting property rights.

The second Brazilian Forest Protection Code was only enacted in 1965, by Law n. 4771/65. This statute regulated several aspects related to the protection of vegetation and established that forests existing in the national territory and other forms of vegetation, recognized as valuable to the lands they cover, are assets of common interest to all inhabitants of the country. It also stated that property rights are to be exercised with legal limitations. By means of an amendment to the Law

n. 4.771/65 in 2001, the Permanent Preservation Area (*Área de Preservação Permanente* or APP) was defined as the area, covered or not by native vegetation, with the environmental function of preserving water resources, the landscape, geological stability, biodiversity, the genetic flow of fauna and flora, protecting the soil, and ensuring the well-being of human populations.[12]

The creation of new Permanent Preservation Areas was also allowed by "acts of public authority" in a broad sense (Law n. 4.771/65, Art. 3). This legal provision was the basis for the issuance of the CONAMA Resolutions n. 302 and 303/2002, which created new hypotheses for APP.

It was in 2012 that the current Forest Protection Code (*Código Florestal*), Law n. 12.651/12, was enacted, having revoked the previous statute (Law n. 4.771/65). This statute is a national law, with application not only to the Federal Government, but also to all subnational entities. The other federative entities (states, federal district, and municipalities) can rule on the protection of forests and vegetation within their territories, according to their peculiarities, but cannot act against the provisions of the Law n. 12.651/12 (Brazilian Constitution, Art. 24, § 1–4).

In 2018, the Supreme Court decided on five lawsuits challenging the constitutionality of the Forest Protection Code.[13] While most of the provisions have been held valid by the Supreme Court, some of them have been declared unconstitutional, and important interpretive lines have been set.

More recently, Law n. 14.285/21 was published, which represented an important change when it provided for the "municipalization" of the rules about APP in marginal strips (protected areas along riverbanks), which will be studied in due course.

12.3.1 Main Features of the 2012 Forest Protection Code

Although we have long had robust legislation to protect forests, the law has not achieved its main purposes of protecting ecosystems, native vegetation, biodiversity, and the sustainable occupation and exploitation of the land. The previous legislations, despite often having relevant protective provisions, have generally not achieved effectiveness in their implementation. And, above all, the former law, especially after its reform in 2001, sharpened the disputes between environmentalists, farmers, businessmen, and politicians.

The current Forest Protection Code arose from the rural sector's dissatisfaction with the former Brazilian forest law, which was very restrictive to agricultural and livestock exploitation and served as the basis for imposing administrative and criminal penalties on offenders. For some commentators, the current statute represented a step back in environmental preservation, by generally reducing the

[12]The amendment was inserted by Provisional Measure (*Medida Provisória*) n. 2.166-67/01.

[13]Brazilian Supreme Court, ADC 42, ADIns 4901, 4902, 4903 e 4937.

protection of native vegetation, facilitating agricultural exploitation, and allowing the consolidation of situations that violated the former law. For others, however, the new Forest Code, although not ideal, has done what was possible to promote sustainability and accommodate environmental, social, and economic demands.

The enactment of Law n. 12.651/12 was marked by one of the most intense political clashes in recent times concerning ecological issues. The political debate involved not only the agribusiness sector and environmental groups, but also academic and research institutions, unions from various sectors, popular movements, different Federal Government departments (advocating divergent positions), and other federative entities. In the National Congress, more than seventy public hearings were held to qualify the social debate around the main modifications related to the regulatory framework of the protection of flora and native vegetation in Brazil.[14] Finally, the Supreme Court considered that the new legislation, in general, is constitutionally valid and holds both institutional and democratic legitimacy.[15]

Like the preceding legislation, the new statute establishes that forests within the national territory, along with other forms of native vegetation deemed beneficial to the lands they cover, are assets of common interest to all inhabitants of the country. It also stipulates that property rights must be exercised within the limits prescribed by general legislation and, in particular, by this law (Law n. 12.651/12, Art. 2). The Forest Protection Code also states that the obligations set forth in this statute have a real right nature (right *in rem* or *propter rem*) and are transmitted to the successor, of any nature, in the event of transfer of ownership or possession of the rural property (Art. 2, § 2).

Law n. 12.651/12 regulates, in summary, the following aspects related to forests and native vegetation: general rules on the protection of vegetation and protected areas; forest exploitation, supply of forest raw material; control of the origin of forest products; control and prevention of forest fires; economic and financial instruments.

In this book, we shall address the following issues due to their importance: the *Permanent Preservation Area* and the *Rural Properties Preservation Area* (*Reserva Legal*),[16] as two of the most important types of protected areas in the country; the control of the use of fire in vegetation; and rules on regularization of activities that were in nonconformity with the previous legislation.

[14] On the disputes underlying the approval of the new Forest Code, see: Fensterseifer and Sarlet (2021), p. 824; Figueiredo (2012), pp. 29–32; Milaré (2018), pp. 1644–1653; Antunes (2013), pp. 1–4.

[15] Brazilian Supreme Court, ADC 42, ADIns 4901, 4902, 4903 e 4937.

[16] The Forest Protection Code mandates a compulsory preservation area on portions of rural properties, referred to in the law as the *Reserva Legal* (or *Legal Reserve*).

12.3.2 Permanent Preservation Areas

The *Permanent Preservation Areas* (*Áreas de Preservação Permanente* or APP) are crucial for protecting the country's ecosystems. Unlike Nature Conservation Units (discussed in the previous chapter), APPs are found in any location that meets the characteristics specified by law. This broad applicability may explain why many conflicts involving the Forest Protection Code are related to the irregular occupation of APPs.

The *Permanent Preservation Area* is defined by the statute as "A protected area, whether covered by native vegetation or not, serving the environmental functions of preserving water resources, landscape, geological stability, and biodiversity; facilitating the genetic flow of fauna and flora; protecting the soil; and ensuring the well-being of human populations" (Law n. 12.651/12, Art. 3, II).

Article 4 of Law n. 12651/12 stipulates that the *Permanent Preservation Areas* provided therein apply to urban and rural areas. In addition, it considers as APP basically the following locations:

(I) the marginal strips of any natural perennial and intermittent watercourse, excluding the ephemeral ones, in differentiated dimensions according to the width of the river;
(II) areas around natural lakes and ponds, also with differentiated widths;
(III) areas surrounding artificial water reservoirs created by the damming or impoundment of natural watercourses, within the strip defined by the environmental license of the project;
(IV) areas surrounding springs and perennial water sources, regardless of their topographical situation, with a minimum protected radius of 50 meters;
(V) slopes, or portions of them, with a gradient steeper than 45°, equivalent to 100% along the line of greatest declivity;
(VI) the *restingas* (sandbanks), as dune fixers or mangrove stabilizers;
(VII) V) the mangroves, in all their extension;
(VIII) the edges of cliffs and plateaus, up to the relief break line, within a strip that is no less than 100 meters in horizontal projection;
(IX) the tops of hills, mountains, and sierras with a minimum height of 100 meters and an average slope greater than 25°, defined by the contour line corresponding to 2/3 (two-thirds) of the minimum height of the elevation relative to its base. The base is determined by the horizontal plane of the nearest plain or adjacent water body, or, in the case of undulating terrain, by the nearest saddle point of the elevation;
(X) areas with an altitude above 1.800 meters, regardless of the type of vegetation present;
(XI) in savanna-like vegetation found on hydromorphic soils (*veredas*), the marginal strip, in horizontal projection, with a minimum width of 50 meters, starting from the permanently marshy and waterlogged space.

The rules of *Permanent Preservation Area* apply to both urban and rural areas. In this regard, the Superior Court of Justice has decided that municipal laws or even the Urban Land Parceling Act (Law n. 6.766/79), which provided for a smaller protection strip for watercourses in certain cases, do not prevail over the rules of the Forest Protection Code.[17]

The imposition of restrictions arising from permanent preservation areas in urban zones, without allowing municipal governments to establish different rules tailored to their particular circumstances, has generated substantial conflicts among politicians, mayors, and business owners on one side, and environmental agencies, associations, and the Public Prosecutor's Office on the other. Eventually, due to pressure from the political and business sectors, the National Congress passed Law n. 14.285/21, which granted municipalities the power to alter the width of APPs on the border of natural waterways in consolidated urban areas. Thus, the mentioned statute introduced amendments to the Forest Protection Code to bring the concept of "consolidated urban area" (Art. 3, XXVI), and to provide that in "consolidated urban areas," municipal (or federal district) law may define marginal strips distinct from those provided under Art. 4, I of Law n. 12.651/12 (Art. 4, § 10 of Law 12.651/12).

The criticism of this new rule is the possibility that municipalities, influenced by political and economic power, may leave urban riverbanks without adequate protection. The Brazilian reality is marked by unplanned urban expansion, with a lack of adequate zoning, planning, and territorial management policies. Clandestine and irregular land developments are proliferating. Political actors, especially at the municipal level, often enticed by the economic interests of construction companies or speculators and pressured by chronic housing deficits, frequently either neglect or even encourage irregular settlements in *Permanent Preservation Areas*. In doing so, they appear to address issues related to a lack of investment, the need for development, municipal revenue growth, or housing shortages, thus potentially securing votes in the next election. However, in the long term, these illegal settlements exacerbate the socio-environmental degradation of cities.[18]

There are two types of *Permanent Preservation Areas*: the *statutory APPs* and the *administrative APPs*. *Statutory APPs* (*APPs legais*) are those directly defined by Law n. 12.651/12 (Art. 4), as described above. *Administrative APPs* (*APPs administrativas*), in turn, are those created by the Executive Branch, based on legal permission.

The previous statute (Law n. 4.771/65, Art. 3) made room for the creation of new APPs (administrative APPs) by means of an "act of public authority" in the broad sense. This legal provision was the basis for the issuance of CONAMA Resolutions n. 302 and 303/2002, which created new hypotheses of APP not directly provided by Law n. 4.771/65. However, the current Forest Protection Code has restricted the

---

[17] Superior Court of Justice, AgRg no REsp 664.886/SC; AREsp 1.312.435/RJ; REsp 1.546.415; REsp 1.770.760/SC.

[18] Moreira (2016).

possibilities of creating *Administrative APPs*, which now can only be done by means of a Decree from the head of the Executive Branch (Law n. 12651/12, Art. 6).

In light of the new legislation, how would CONAMA Resolutions n. 302 and 303/02 stand, which, as mentioned above, created new hypotheses for *Administrative APPs*? Remember that these regulations were issued by CONAMA, and not by the President of the Republic (head of the Federal Executive Branch). This was possible under the previous statute but is no longer possible under the current one.

This was the subject of CONAMA's deliberation itself, which decided, in 2020, to revoke these regulations (through Resolution n. 500/2020). To do so, CONAMA basically stated that it would no longer have power to create APPs based on the new legislation, and that the New Forest Protection Code had exhausted the matter, which would generate a *tacit revocation* of these regulations.

This revocation by CONAMA was the target of judicial review before the Supreme Court. Eventually, the Supreme Court ruled that the revocation of the Resolutions n. 302 and 303 were contrary to the Constitution. The Court reasoned that there was an "unconstitutional deregulation," when it stated that the "revocation of operational rules that set measurable parameters necessary for compliance with environmental legislation, without replacing or updating them, undermines compliance with the Constitution, current legislation, and international commitments." It considered that there was an apparent dismantling of the administrative framework of environmental protection, with violation of the precautionary principle and the principle of non-regression.[19]

Therefore, currently, in addition to the APPs provided for in the Forest Protection Code, we also have the *Administrative APPs* provided for in the CONAMA Resolutions n. 302 and 303/02. A presidential Decree creating *Administrative APPs* based on Art. 6 of Law n. 12651/12 has not yet been issued.

The importance of studying *Permanent Preservation Areas* arises from their strict protective status and their widespread presence across the country. This results in numerous conflicts among landowners, farmers, politicians, environmental agencies, environmental organizations, and public prosecutors.

Under Art. 7 of Law n. 12.651/12, vegetation located in *Permanent Preservation Areas* must be maintained by the owner of the land, establishing a general duty to preserve vegetation in these areas regardless of ownership. If vegetation has been removed, the owner, possessor, or any titleholder is obligated to restore it, except for uses authorized by law (Art. 7, § 1). This obligation has a *propter rem* nature, meaning it is attached to the property and transfers to the successor upon any transfer of ownership or possession (Art. 7, § 2).

Superior Court of Justice precedents also classify Permanent Preservation Areas as *non-buildable areas* (*área non aedificandi*), requiring the restoration of any illegally degraded land.[20] Moreover, unauthorized construction in these areas is deemed an environmental crime (Law n. 9.605/98, Art. 64).

[19] Brazilian Supreme Court, ADPF 747, 748, 749.

[20] Superior Court of Justice, AgRg no REsp 1.367.968/SP; REsp 1394025/MS; REsp 1344525/SC.

Unlawful occupation and construction in *Permanent Preservation Areas* is a common issue in environmental lawsuits in Brazil. Generally, environmental agencies or courts must order the evacuation of the site, demolition of the building, debris removal, and restoration of the degraded area. Furthermore, the Superior Court of Justice rejects the application of the proportionality principle to legitimize any environmental damage that has become established over time. In other words, regardless of how long an APP has been occupied, if the occupation is illegal, the removal of occupants and buildings must be ordered.[21] Similarly, the Superior Court of Justice has ruled that illegal occupation of a *Permanent Preservation Area* must cease, regardless of the presence of other unlawful occupations or the area's population density.[22]

Pursuant to Art. 8 of the current Forest Protection Code, intervention or removal of native vegetation in Permanent Preservation Areas is permitted only in cases of public utility, social interest, or low environmental impact. However, the removal of native vegetation that protects springs, dunes, and sandbanks may only be authorized for reasons of public utility (Art. 8, § 1).

The cases of public utility, social interest, and low environmental impact are described under Art. 3, VIII, IX, and X of Law n. 12.651/12. Examples can be given: (a) as of *public utility* the cases of national security and sanitary protection activities, and certain infrastructure works; (b) as of *social interest* the sustainable agroforestry exploitation practiced in a small property or family rural possession or by traditional groups of people and communities, and land regularization of human settlements occupied predominantly by low-income population in consolidated urban areas; (c) as of *low environmental impact* the opening of small internal access ways and their bridges, when necessary for the crossing of a water course, the access of people and animals to obtain water or the removal of products originating from sustainable agroforestry management activities, and the implantation of trails for the development of ecotourism.

In any case, environmental agency permission is not required for the emergency execution of national security activities and civil defense works aimed at preventing and mitigating accidents in urban areas (Art. 8, § 3).

In several judicial review lawsuits, the Supreme Court has reduced the possibilities of intervention in *Permanent Preservation Areas*. Accordingly, the Court deemed unconstitutional the insertion of two cases of *public utility* by Law n. 12.651/12: waste management activities and the implementation of facilities necessary to hold sporting competitions. This is because, for both cases, it is possible to find alternative locations, and it is unjustified to maintain such activities in an environmentally relevant area. Furthermore, in relation to waste management, because of the environmental impacts, especially on groundwater, it seems incompatible with these protected areas.

[21] Superior Court of Justice, AgInt no REsp 1.542.756/SC.

[22] Superior Court of Justice, REsp 1457851/RN; REsp 769.753/SC; REsp 1782692/PB.

The Brazilian Supreme Court also introduced an additional requirement for interventions in *Permanent Preservation Areas*. It stipulates that, in all cases of public utility or social interest, the absence of a technical or locational alternative must be demonstrated. In other words, it must be proven that the only feasible location for the proposed work or activity is within that specific environmentally protected area.[23]

12.3.3 Rural Property Preservation Area

The Forest Protection Code prescribes that a portion of each rural property must be preserved with native vegetation, known as *Reserva Legal* (*Legal Reserve*) or *Rural Property Preservation Area*, where deforestation is prohibited. This is the protected portion of a farm or ranch and is mandatory in nearly all cases.

Under Art. 3, III of Law n. 12.651/12, the *Rural Property Preservation Area* (*Reserva Legal*) is the area located within a rural property or possession, delimited in accordance with Art. 12, with the function of ensuring the sustainable economic use of the natural resources of the rural property, supporting the ecological processes conservation and rehabilitation, and promoting the biodiversity conservation, as well as the shelter and protection of wild fauna and native flora.

Article 12 establishes that every rural property must maintain an area with native vegetation coverage, in addition to any *Permanent Preservation Areas*, observing the following minimum percentages relative to the total property area: I) for properties within the Legal Amazon,[24] the *Rural Property Preservation Area* must correspond to 80% of the property located in forest areas, 35% in *cerrado* (savanna) areas,[25] and 20% in general grassland areas; II) for properties located in other regions of the country, the area to be preserved must correspond to 20% of the property.

It should be noted that every rural property must be registered in the *Rural Environmental Register* (*Cadastro Ambiental Rural* or CAR), and the *Rural Property Preservation Area* must also be documented within this registration. The *Rural Environmental Registry*, created by the New Forest Protection Code, is the national electronic public registry, mandatory for all rural properties, to integrate environmental information from rural properties and possessions, compose a database for control, monitoring, environmental and economic planning, and combating deforestation (Law n. 12.651/12, Art. 29). The absence of the property's inscription in this

[23] Brazilian Supreme Court, ADC 42, ADI 4901, 4902, 4903 e 4937.

[24] Brazil's Legal Amazon (*Amazônia Legal*) is the largest socio-geographic division in Brazil, containing all nine states in the Amazon basin. The government designated this region in 1948 based on its studies on how to plan the economic and social development of the Amazon region.

[25] The *Cerrado* is a vast ecoregion of tropical savanna in eastern Brazil. The main habitat types of the *Cerrado* consist of forest savanna, wooded savanna, park savanna and gramineous-woody savanna. The *Cerrado* also includes savanna wetlands and gallery forests.

registry, for example, blocks the suppression of vegetation for agricultural exploration (Art. 12, § 3), as well as the concession of agricultural financing (Art. 78-A).

An important mechanism, present in the previous statute but strengthened in the current legislation, is the *compensation of Rural Property Preservation Area* (*compensação de reserva legal*) between different properties. This technique resembles the cap-and-trade system in the carbon market. Law n. 12.651/12 allows a property with a preserved area smaller than the legally required minimum to compensate by maintaining a larger preserved area on another property. A landowner with a preserved area exceeding the minimum requirement can use the surplus for purposes such as establishing an *Environmental Reserve Quota* (*Cota de Reserva Ambiental*) (Art. 13, § 1; Art. 15, § 2).

The *Environmental Reserve Quota* is a document that represents the area with native vegetation (Art. 44). This document, similar to a bond, can be traded or sold to a landowner who, for instance, wants to exploit a higher percentage of his land than allowed under the Forest Protection Code. The *Environmental Reserve Quota* can only be used to compensate for the preserved area of a rural property located within the same *biome*[26] as the area to which the Quota is linked (Law n. 12.651/12, Art. 48, § 2). Consequently, this has established a genuine "market of preserved areas."

The Attorney-General of the Republic (*Procurador Geral da República*), in a judicial review action, sought a declaration of unconstitutionality for the rules allowing certain properties to maintain a smaller protected area. However, the Supreme Court upheld the constitutionality of these provisions, stating that the legislation employs innovative incentive mechanisms to promote environmental protection, moving beyond the traditional—and often ineffective—*command-and-control* approach. The Court noted that those challenging the New Forest Protection Code's constitutionality essentially want the Court to impose *command-and-control* as the sole model of environmental legislation permitted in Brazil, excluding *market-based* approaches. Additionally, the Court remarked that opposition to market solutions runs counter to global trends in environmental policy, citing the success of international carbon emissions trading as an example. The Court added only one restriction: the two properties used for compensation must share an *ecological identity*.[27]

Regarding the protection framework of the rural property preservation area, the law requires that it must be preserved with the coverage of native vegetation (Law n. 12.651/12, Art. 17). Exceptionally, economic use of this area is permitted through *sustainable management* (Art. 17, § 1).

---

[26] Brazil has six distinct natural regions, called *biomes*: the Amazon, Atlantic Forest, *Cerrado*, *Pantanal*, *Pampas* and *Caatinga*. Each of these biomes have their own peculiarities, located in different climatic zones and hosting a wide variety of animal and plant species. More about Brazilian biomes in: Global Change Ecology (2022).

[27] Brazilian Supreme Court, ADC 42, ADI 4901, 4902, 4903 e 4937.

12.3.4 Regulating the Use of Fire in Farming Activities

The use of fire is a common tool in agriculture. Fire brings a series of facilities and benefits to farmers since it prepares the land for cultivation and is cheap and accessible. However, fire is associated with many environmental impacts. In the Amazon, the problem is even more serious, because the use of fire is cultural and for many isolated small farmers it is the only option, since it is not possible to have access to new technologies and tools. Because of this, the volume of wildfires in the Amazon and other Brazilian regions has reached record highs.

The use of fire is regulated primarily by the Forest Protection Code. As a general rule, the use of fire in vegetation is forbidden (Art. 38). However, Law n. 12.651/12 provides exceptions, allowing: (a) the use of fire in areas where specific conditions justify its use in agricultural or forestry practices, with prior approval from the state environmental agency; (b) controlled burning in Conservation Units, in accordance with the respective management plan and upon prior approval of the Conservation Unit's managing body; (c) activities for scientific research as part of an approved research project conducted by a recognized institution, upon prior approval by the environmental agency; (d) fire prevention and firefighting practices; and (e) the use of fire in subsistence agriculture performed by traditional and Indigenous populations (Art. 38, I–III and § 2).

It is pertinent to highlight the recent Law n. 14.944/24, which addresses the National Policy for Integrated Fire Management. This new statute establishes guidelines for the use of fire in rural areas, with a focus on sustainability and biodiversity protection. The legislation aims to gradually replace the practice of burning with alternative techniques, particularly within traditional and Indigenous communities that have practices related to fire management. This law—which amends the Forest Protection Code and the Environmental Crimes Law (Law n. 9.605/98)—allows Indigenous and *Quilombola* communities to conduct controlled burns for subsistence agriculture, provided specific conditions are met, such as prior agreements with the resident community and notification to the forest firefighting teams responsible for the area. These activities must be planned to take place during appropriate seasons and with adequate safety measures. The implementation of this policy will be coordinated by IBAMA,[28] in partnership with FUNAI,[29] the Palmares Cultural Foundation,[30] and other agencies.

[28] About IBAMA (Brazilian Institute for the Environment and Renewable Natural Resources), consult Sect. 7.2.2.

[29] FUNAI (National Indian Foundation or *Fundação Nacional do Índio*) is a federal agency responsible for implementing Indigenous protection policies, as established by Law n. 5.371/67.

[30] The creation of the Palmares Cultural Foundation (*Fundação Cultural Palmares*) was authorized by Law n. 7.668/88, with the purpose of promoting the preservation of the cultural, social, and economic values resulting from the black influence in the formation of Brazilian society. The expression *quilombolas* refers, in short, to the descendants and remainders of communities formed by runaway slaves (the *quilombos*), between the sixteenth century and 1888 (when slavery was abolished), in Brazil.

This new law provides for permitted uses of fire in a manner similar to the Forest Code, with some innovations. In addition to the previously established cases, Law n. 14.944/24 permits the use of fire in the following circumstances (Art. 30): (a) in cultural practices and subsistence agriculture conducted by Indigenous peoples, *Quilombola* communities, other traditional communities, and family farmers, in accordance with their customs and practices; (b) in the training and development of forest firefighting brigades; (c) in the harvesting of sugarcane, as a method for defoliation and facilitation in areas that cannot be mechanized, in accordance with regulations set by the competent state authority.

A controversial issue concerns the burning of sugarcane straw in agro-industrial activities. Despite its significant environmental impacts, particularly on the climate, this activity is permitted by law, as decided by the Superior Court of Justice. However, it requires specific authorization from the environmental agency, including licensing and a prior Environmental Impact Study.[31]

Similarly, in a landmark case involving a conflict between a municipal law prohibiting the burning of sugarcane straw and a state law allowing gradual elimination, the Supreme Court ruled in favor of the state law. The Court reasoned that this activity is a matter of regional, rather than predominantly local, interest.[32] In a critical analysis of this decision, Wedy pointed out that this Supreme Court's ruling did not mention either the severe risk of greenhouse emissions deriving from the fires, or the Climate Change National Policy Act, and was against international treaties, like the Paris Agreement.[33] In any case, what we could observe is a solution given by the Supreme Court to accommodate diverging interests—the economic interest of a region, highly dependent on the sugarcane straw burning, and the collective interest in clean air and climate balance.

Accordingly, it becomes clear that the recently enacted Law n. 14.944/24 has embraced the courts' interpretation regarding the permissibility of using fire for burning sugarcane straw.

12.3.5 Regularization of Non-Compliant Activities

As highlighted earlier, Brazil had for a long time a well-drafted Forest Protection Code, at least since 1965, with provisions for rural property preservation areas and Permanent Preservation Areas. However, the law "didn't stick," and non-compliance with the rules for protection of these ecologically relevant areas became widespread for many years, throughout the national territory.

The New Forest Protection Code (Law n. 12.651/12) significantly reduced the extent of *Rural Property Preservation Areas* and *Permanent Preservation Areas* and

[31] Superior Court of Justice, EREsp 418.565/SP; REsp 1668060/SP.

[32] Brazilian Supreme Court, RE 586224.

[33] Wedy (2017).

allowed for more cases of intervention in these areas. At the same time, to bring landowners back into compliance with environmental legislation, the new statute introduced provisions for the regularization and restoration of deforested areas, offering exemption from penalties as a reward for those who successfully regularize their situation. This initiative is known as the *Environmental Regularization Program* (*Programa de Regularização Ambiental*).

The legislator designated July 22, 2008,[34] as the "watershed" date determining eligibility for regularization under the new statute. In other words, individuals who illegally deforested *Rural Property Preservation Areas* and *Permanent Preservation Areas* before this date could benefit from the new provisions to avoid administrative and criminal penalties, provided they commit to restoring vegetation in these areas and, importantly, fulfill these commitments.[35]

The Forest Protection Code also set forth different treatments for illegal deforestation in these areas, distinguishing between actions occurring before and after July 22, 2008. For unauthorized vegetation removal in Permanent Preservation Areas after this date, the Code prohibits issuing new authorizations for vegetation suppression until vegetation recovery obligations are met.[36] In other words, a *contrario sensu*, deforestation that took place before this date would not preclude new authorizations for vegetation suppression. Similarly, the Code demands the immediate suspension of activities in *Rural Property Preservation Areas* unlawfully deforested after July 22, 2008.[37] Again, this implies that illegal deforestation occurring before this date does not require the immediate suspension of activities in these areas.

The *Environmental Regularization Program* has faced significant criticism, with claims that the new legislation rewards lawbreakers, encourages deforestation, and discourages those who seek to comply with environmental laws. The Supreme Court, however, upheld the validity of these provisions.[38] Judges emphasized that Law n. 12.651/12 introduced mechanisms to encourage violators to repair the damage they caused, thus restoring their compliance with the law, especially considering the widespread legal violations that had persisted until then.

References

Antunes PB (2013) Comentários ao Novo Código Florestal. Atlas, São Paulo
Fensterseifer T, Sarlet IW (2021) Curso de Direito Ambiental. 2. ed. Forense, Rio de Janeiro

[34]The date July 22, 2008, corresponds to the day Decree n. 6.514/08, on federal environmental administrative proceedings, was promulgated. Despite being an "arbitrary" date and unrelated to Law n. 12.651/12, it was held constitutional by the Supreme Court (ADC 42, ADIns 4901, 4902, 4903 and 4937).

[35]Law n. 12.651/12, Art. 59–68.

[36]Law n. 12.651/12, Art. 7, § 4.

[37]Law n. 12.651/12, Art. 17, § 3.

[38]Brazilian Supreme Court, ADC 42, ADIns 4901, 4902, 4903, and 4937.

Figueiredo GJP (2012) In: Milaré É, Machado PAL (eds) Novo Código Florestal. Revista dos Tribunais, São Paulo

Global Change Ecology (2022) Brazilian biomes: intro series. https://globalchangeecology.com/2022/06/30/brazilian-biomes-intro-series/. Accessed 18 Nov 2024

Greenpeace (2024) Forests Brazil and the Amazon Forest. https://www.greenpeace.org/usa/issues/brazil-and-the-amazon-forest/. Accessed 18 Nov 2024

Milaré É (2018) *Direito do Ambiente*. 11. ed. Revista dos Tribunais, São Paulo

Moreira RMC (2016) Áreas de preservação permanente e urbanização: muitos conflitos e uma controvérsia. Interesse Público 18(99):90–116

United Nations Environment Programme (2020) Convention on Biological Diversity. Aichi Target 11. https://www.cbd.int/aichi-targets/target/11. Accessed 18 Nov 2024

Wedy G (2017) Climate legislation and litigation in Brazil. Sabin Center for Climate Change Law. Columbia Law School, October 2017. https://climate.law.columbia.edu/sites/default/files/content/Wedy-2017-10-Climate-Legislation-and-Litigation-in-Brazil.pdf. Accessed 18 Nov 2024

Chapter 13
Climate Change Law and Climate Change Litigation

13.1 Introduction

Since it was first mentioned in the early nineteenth century and developed in the second half of the twentieth century, global warming and climate change have remained a topic of discussion, and it has been a debatable issue among politicians and the scientific community. By the same token, it has been the most controversial environmental problem facing the world, and it has become an issue of high priority on international political agendas.[1] Unlike the transboundary pollution that first affects countries downwind or downriver, atmosphere pollution threatens to cause serious damage to the entire planet.[2] To add complexity to this debate, this is a problem that is still invisible to most people (although this invisibility is waning in the face of increasing climatic disturbances), and most serious consequences are expected to be felt in the long term. The issue is gaining momentum within the international community, with demonstrations drawing widespread attention to the climate crisis, as evidenced by actions from the youth-led movement *Fridays for Future* and, since 2021, the more radical actions of the *Last Generation* organization.

The following section addresses how discussions on climate policy began, the international community's response, and how Brazilian law has approached this crisis.

[1] Berlie (2018).

[2] Leape et al. (2018), p. 2073.

R. Martins Costa Moreira, G. Wedy, *Brazilian Environmental and Climate Change Law*, https://doi.org/10.1007/978-3-031-87998-2_13

13.2 The Awakening to a New Crisis: The Green House Effect

The issue of climate change and, in general, the impacts caused by atmospheric pollution are not necessarily new. New are the heightened debates about the problem and how seriously governments and the international community have been reacting to the possible consequences of global warming.

Studies on climate change already existed in the nineteenth century. The central question is about how climate change went from a subject of scientific inquiry, which engaged only a few scientists, to one of the dominating issues of our age.[3] Indeed, before the main concern became tackling global warming, however, it had been the ozone hole that had captured the attention of countries and international organizations.

As well summarized by Yergin, in the 1970s and especially in the 1980s, it was definitively proven that "a 'hole' was opening up in the ozone over Antarctica. The chlorofluorocarbons were eating at the ozone, literally thinning out and depleting the layer in the atmosphere." Because ozone absorbs ultraviolet radiation, its loss could be devastating to animal and plant life and would threaten to cause a truly worldwide epidemic of skin cancer. "Such was the fear that in record time – by 1987 – some twenty-four countries signed on to the Montreal Protocol, which would restrict chlorofluorocarbons." The Montreal Protocol has had an undeniable influence on climate change policy, not least by showing that "countries could come together quickly and agreed to eliminate a common environmental threat."[4]

The climate change problem, however, has proved to be far more challenging than ozone depletion.[5] As Yergin observed, the relevant universe was so much smaller. "Fewer than forty companies manufactured chlorofluorocarbons, and just two had half the market. But the whole world burned fossil fuels. Nevertheless, global warming, with all its complexity, was by the summer of 1988 entering the political arena."[6]

Human-driven global warming is, in all its aspects, highly controversial. Even for the skeptics it is an empirical fact that the Earth's climate has been getting warmer since at least the past century. And yet, for a minority of scholars and scientists, we do not know how much humans have contributed to this warming and if human-caused emissions of greenhouse gases are the prevailing reason for the most recent warming, compared to natural causes.[7] So-called *climate deniers*—those who reject the idea that anthropogenic emissions are causing global warming and climate disruptions—are often influenced by the dissemination of information and studies

[3] Yergin (2012), p. 425.
[4] Yergin (2012), pp. 459–460.
[5] Leape et al. (2018), p. 2073.
[6] Yergin (2012), p. 460.
[7] Curry (2019), p. 89.

commissioned by those with the most to lose from the transition to renewable energy: the oil and gas industry.[8]

However, the prevailing view is that global warming is unequivocally human-driven, as the IPCC has widely stated. For most scientists, as a result of the buildup of carbon dioxide and other gases in the atmosphere, the Earth is on the verge of unprecedented global climate change. But despite a growing consensus that some global warming will occur, there is also great uncertainty concerning the severity of its consequences.[9]

To reach an international consensus on climate change, two United Nations agencies—the World Meteorological Organization and UNEP—organized an Intergovernmental Panel on Climate Change (IPCC) in 1988. The IPCC is the international body for assessing the science related to climate change. It provides policymakers with regular assessments of the scientific basis of climate change, its impacts and future risks, and options for adaptation and mitigation. It is composed of nearly 2500 independent scientists from more than 100 countries who are charged with reviewing and summarizing what is known about climate change.

In 1990, the IPCC released its first report summarizing scientific knowledge about climate change. The IPCC First Assessment Report (also called "AR1") found that it is "certain" that "there is a natural greenhouse effect which already keeps the Earth warmer than it would otherwise be," and that "emissions resulting from human activities are substantially increasing the atmospheric concentrations of the greenhouse gases carbon dioxide, methane, chlorofluorocarbons (CFCs), and nitrous oxide. These increases will enhance the greenhouse effect, resulting on average in an additional warming of the Earth's surface."[10]

The consensus has been growing as new studies and reports have been carried out. The sixth and last Report (AR6), issued in 2022, asserted that "it is unequivocal that human influence has warmed the atmosphere, ocean and land. Widespread and rapid changes in the atmosphere, ocean, cryosphere and biosphere have occurred. (…) Human-induced climate change is already affecting many weather and climate extremes in every region across the globe. Evidence of observed changes in extremes such as heatwaves, heavy precipitation, droughts, and tropical cyclones, and, in particular, their attribution to human influence, has strengthened since AR5. (…) Climate change is already affecting every inhabited region across the globe, with human influence contributing to many observed changes in weather and climate extremes. (…) Global surface temperature will continue to increase until at least mid-century under all emissions scenarios considered. Global warming of 1.5°C and

[8]With regard to the scientific consensus established in the peer-reviewed literature on the severity of the harmful effects caused by human-induced global warming—and the attempts to distort and undermine science in favor of political, ideological, and economic interests—see: Conway and Oreskes (2011) and Oreskes (2019).

[9]Leape et al. (2018), p. 2089.

[10]Intergovernmental Panel on Climate Change (IPCC) (1990).

2°C will be exceeded during the 21st century unless deep reductions in CO2 and other greenhouse gas emissions occur in the coming decades."[11]

IPCC reports and scientific findings have influenced subsequent international agreements to tackle global warming, which will be addressed in the next chapter. Two key issues have been proven to be relevant to steer actions at international and national levels. First, the issue of *climate justice*: the uneven impacts of global warming, in which the least developed countries and communities, and therefore those that contribute least to climate change, are those that suffer most from its consequences. Second, the precautionary principle: faced with uncertainty, especially regarding the consequences of global warming, the adoption of precautionary measures takes on a prominent place to avoid probable and irreversible damage.

13.3 The International Response to the Climate Crisis

The climate crisis was initially addressed at the international level. An international response to global warming began with a UN General Assembly resolution in December 1988 declaring climate change a "common concern of mankind" and calling for global action to tackle the problem. At the end of 1990, the General Assembly adopted a resolution called Protection of Global Climate for Present and Future Generations of Mankind that established a process for the negotiation of an international framework convention on climate change.[12]

On June 12, 1992, 154 other nations attending the Rio Earth Summit signed a Framework Convention on Climate Change (UNFCCC). It can be considered as the "birth" of International Climate Change Law. This treaty was thought to become a framework for international cooperation against climate change by limiting average global temperature increase and the resulting climate change, and coping with impacts that were, by then, inevitable.[13]

The Convention was adopted on 9 May 1992 and opened for signature a month later at the UN Conference on Environment and Development in Rio de Janeiro, Brazil. It entered into effect on March 21, 1994, after receiving 50 ratifications. The Convention has now 186 Parties and is approaching universal membership. Since the adoption of the Convention, Parties have continued to negotiate in order to agree on decisions and conclusions that will advance its implementation. They have done so, since the Convention entered into force, in the Conference of the Parties (COP) and its subsidiary bodies. In addition to this "routine" work on advancing the implementation of the Convention, Parties launched a new round of negotiations at COP1 (Berlin, March/April 1995) to strengthen the commitments of *Annex I*

[11] Intergovernmental Panel on Climate Change (IPCC) (2021).

[12] Leape et al. (2018), p. 2098.

[13] United Nations Climate Change (2024c).

Parties.[14] These negotiations resulted in the adoption of the Kyoto Protocol at COP3 (Kyoto, December 1997).[15]

Since the beginning of the Conferences of the Parties, the "conflict" between developed and developing countries has become clear. Some 75% of total accumulated emissions of CO2 between 1860 and 1990 had come from industrialized nations. But they amount to only 20% of the world's population. In Yergin's words, "as carbon limits seemed to take on greater likelihood, the developing countries became more vociferous in opposing limits on their use of hydrocarbons and the constraints such limits could impose on their economic growth." This clash between developed and developing nations was a dominant issue when national delegations convened in Berlin in 1995. The Berlin Mandate concluded that while the industrial nations would take on specific targets in the next phase of global climate regulation, the developing countries would be spared from such obligations. That *differentiated responsibility* would come to be an increasingly more important battlefield therein the global politics of climate change.[16]

This controversy continued at COP3 in Kyoto. First, there was a conflict between Europe and United States. The Europeans wanted the Americans to take deeper cuts. The United States refused. Finally, with the decisive contribution of then Vice President Al Gore, the deadlock was broken, with the result that the United States, Europe, and Japan all ended up with roughly the same binding targets—CO2 emissions between 6% and 8% lower by 2008—12 compared with 1990.[17]

The main question at Kyoto was the debate as to whether the developing nations would also make binding commitments about reducing emissions. The core of the dispute, as stated by Yergin, is that if "one looked backward at emissions, then the developing countries had a strong case. But, if one looked forward, the developing countries would be responsible for a growing share of CO2 as their economies grew."[18] But without binding targets for developing countries, it would be very hard to turn an agreement at Kyoto into a treaty approved by the U.S. Senate. Within the United States, the approval of the Kyoto Protocol encountered a major obstacle with the so-called "Byrd-Hagel Resolution," which said that any climate agreement should maintain U.S. competitiveness and that all the major emitters should be included—including the developing world. However, Kyoto did not include them, so the U.S. could not ratify the protocol.[19]

---

[14] Annex I Parties includes the industrialized countries that were members of the OECD (Organization for Economic Co-operation and Development) in 1992, plus countries with economies in transition (the "EIT Parties"), including the Russian Federation, the Baltic States, and several Central and Eastern European States.

[15] United Nations Climate Change (2024a).

[16] Yergin (2012), p. 485.

[17] Yergin (2012), p. 488.

[18] Yergin (2012), p. 489.

[19] Yergin (2012), p. 491.

It is important to notice that Brazil, because of its non-inclusion in the mentioned *Annex 1 Parties*, has not an emission reduction or any limitation commitments under the Kyoto Protocol.

One key aspect of the Kyoto Protocol was the establishment of flexible market mechanisms, which are based on the trade of emissions permits. Under the Protocol, countries must meet their targets primarily through national measures. However, the Protocol also offers them additional means to meet their targets by way of three market-based mechanisms. Thus, three flexibility and implementation mechanisms were devised: International Emissions Trading; Clean Development Mechanism (CDM); and Joint implementation (JI).[20]

The protocol established greenhouse gas emissions as a new commodity. Parties with commitments under the Kyoto Protocol (*Annex B Parties*) have accepted targets for limiting or reducing emissions. These targets are expressed as levels of allowed emissions or assigned amounts along the 2008–2012 commitment period. The allowed emissions are divided into assigned amount units (AAUs).

Emissions trading, as set out under Article 17 of the Kyoto Protocol, allows countries that have emission units to spare—emissions permitted to them but not "used"—to sell this excess capacity to countries that are over their targets. Thus, a new commodity has been created in the form of emission reductions or removals. Since carbon dioxide is the principal greenhouse gas, people speak simply of trading in carbon. Carbon is now tracked and traded like any other commodity. This is known as the *carbon market*. As we shall see, although the carbon market is booming in some countries, mainly in Europe, in Brazil this market is still incipient.[21]

Another innovative tool of the Kyoto Protocol was the Clean Development Mechanism (CDM). The CDM, defined under Article 12 of the Protocol, allows a country with an emission-reduction or emission-limitation commitment under the Kyoto Protocol to implement an emission-reduction project in developing countries. Such projects can earn saleable Certified Emission Reduction (CER) credits, each equivalent to one ton of CO2, which can be counted towards meeting Kyoto targets. A CDM project activity might involve, for example, a rural electrification project using solar panels or the installation of more energy-efficient boilers. The mechanism stimulates sustainable development and emission reductions, while giving industrialized countries some flexibility in how they meet their emission reduction or limitation targets.

A CDM project must provide emission reductions that are additional to what would otherwise have occurred. The projects must qualify under the assessment of a rigorous public registration issuance process. Approval is given by Designated National Authorities. Public funding for CDM project activities must not result in the diversion of official development assistance. The mechanism is overseen by the CDM Executive Board, answerable ultimately to the countries that have ratified the

[20]United Nations Climate Change (2024h)?

[21]United Nations Climate Change (2024b).

Kyoto Protocol. Operational since the beginning of 2006, the mechanism has already registered more than 1650 projects and is anticipated to produce CERs (certified emission reduction (CER) credits) amounting to more than 2.9 billion tons of CO2 equivalent in the first commitment period of the Kyoto Protocol, 2008–2012.[22]

Brazil has actively participated in the Clean Development Mechanism. In November 2004, Brazil registered the first CDM project, becoming a pioneer in this program. This was the *NovaGerar* project, developed in the city of Nova Iguaçu in Rio de Janeiro, which includes investments in a gas collection system and modular electricity generation plants at a former open dump and a new sanitary landfill. The project attracted the interest of the Dutch Government and the World Bank for carbon credits purchase, based on the CDM. Registered in the Clean Development Mechanism, the project issued emissions reduction certificates, which were negotiated by the Dutch Clean Development Mechanism Fund and purchased by industrialized countries that failed to meet their pollutant reduction targets under the Kyoto Protocol.[23] Since then, until September 2019, Brazil has had 343 projects registered with the UNFCCC, representing 5% of the world total of 7808 projects, being the third country with the most registered projects, only behind China and India.[24] Currently, however, the acceptance of requests for the issuance of letters of approval for CDM projects in Brazil is suspended to allow for the transition to the Article 6.4 mechanism under the Paris Agreement.

Finally, the mechanism known as *Joint Implementation* allows a country with an emission reduction or limitation commitment under the Kyoto Protocol (Annex B Party) to earn emission reduction units from an emission-reduction or emission removal project from another Annex B Party, each equivalent to one ton of CO2, which can be counted towards meeting its Kyoto target. A joint implementation project must provide a reduction in emissions by sources, or an enhancement of removals by sinks, that is additional to what would otherwise have occurred. Projects must have approval of the host Party and participants must be authorized to participate by a Party involved in the project.[25] Joint implementation offers Parties flexible and cost-efficient means of fulfilling part of their Kyoto commitments, while the host Party benefits from foreign investment and technology transfer. Brazil is not part of this mechanism, as it is included in the so-called "Annex B."

A few years had passed since the Kyoto Protocol, with several conferences among the parties, until the COP21 in Paris in 2015, which resulted in the *Paris Agreement*. The Paris Agreement is a legally binding international treaty on climate change. It was adopted by 196 Parties and entered into force on November 4, 2016. Its goal is to limit global warming to well below 2 °C, preferably to 1.5 °C, compared to pre-industrial levels. To achieve this long-term temperature goal, countries aim to reach global peaking of greenhouse gas emissions as soon as possible to achieve a

[22] United Nations Climate Change (2024f).

[23] United Nations Climate Change (2022), World Bank (2004) and Ribeiro and de Souza (2009).

[24] Leite and Fernandes (2021).

[25] United Nations Climate Change (2024d).

climate-neutral world by mid-century. The Paris Agreement is a landmark in the multilateral climate change process because, for the first time, a binding agreement has brought all nations together into a common cause to undertake ambitious efforts to tackle climate change and adapt to its effects.

The implementation of the Paris Agreement requires economic and social transformation, based on the best available science. The Paris Agreement works on a 5-year cycle of increasingly ambitious climate action carried out by countries. By 2020, countries have submitted their plans for climate action known as nationally determined contributions (NDCs). In their NDCs, countries have communicated actions they will take to reduce their Greenhouse Gas emissions to reach the goals of the Paris Agreement. Countries also communicate in the NDCs actions they will take to build resilience to adapt to the impacts.[26]

Nationally determined contributions are the heart of the Paris Agreement and the achievement of its long-term goals. NDCs embody efforts by each country to reduce national emissions and adapt to the impacts of climate change. The Paris Agreement requires each Party to prepare, communicate, and maintain successive nationally determined contributions that they intend to reach. Parties shall pursue domestic mitigation measures, to achieve the objectives of such contributions.[27]

Brazil communicated to the Secretariat of the United Nations Framework Convention on Climate Change (UNFCCC) its nationally determined contribution (NDC), updated in the context of the Glasgow Climate Pact (COP26, in 2021). Through this communication, Brazil confirms its commitment to reduce its greenhouse gas emissions in 2025 by 37%, in comparison with 2005. Additionally, Brazil commits to reduce its emissions in 2030 by 43%, in comparison with 2005. Brazil's commitments also include a long-term objective to achieve climate neutrality by 2050.[28] Brazil's most recent Nationally Determined Contribution (NDC) was presented at the 2024 United Nations Climate Change Conference (COP29) in Baku, Azerbaijan. The country committed to a reduction of 59–67% by 2035.

Another program of great importance for stabilizing greenhouse gases in the atmosphere is the "Reducing Emissions from Deforestation and Forest Degradation in Developing Countries" (REDD). Forests have been recognized as important carbon sinks because they absorb CO2 in the chemical process of photosynthesis, retaining carbon in the tree trunks, and releasing oxygen. Thus, they contribute greatly to the removal of CO2 from the atmosphere.

REDD, in its origin, aims to help realize forest solutions to the climate emergency by avoiding carbon emissions and fostering carbon sequestration. In pursuing this international goal, REDD promotes approaches that ensure the environmental integrity of carbon emissions reductions while supporting non-carbon benefits—from

[26] United Nations Climate Change (2024g).

[27] United Nations Climate Change (2024e).

[28] United Nations Climate Change (2021).

safeguarding biodiversity to supporting local livelihoods and promoting the rights of Indigenous peoples.[29]

The idea of REDD became formal at the UNFCCC 13th Conference of the Parties (COP13) in Bali in 2007. After discussions among the members, at COP16 (2010) in Cancun, REDD was renamed *REDD+* to reflect the new components of the greenhouse gas emissions mechanism. *REDD+* is defined as "reducing emissions from deforestation and forest degradation in developing countries, and the role of conservation, sustainable management of forests and enhancement of forest carbon stocks." *REDD+* aims to incentivize developing countries to contribute to climate change mitigation actions in the forest sector by means of: reducing carbon emissions from deforestation; reducing carbon emissions from forest degradation; conservation of forest carbon stocks; sustainable management of forests; and enhancement of forest carbon stocks.[30] *REDD+* goes beyond simply deforestation and forest degradation and includes the role of conservation, sustainable management of forests, and enhancement of forest carbon stocks. The framework is commonly referred to as the *Warsaw Framework for REDD+* adopted at COP19 in Warsaw, in December 2013 and provides the complete methodological and financing guidance for the implementation of *REDD+* activities. *REDD+* is also recognized in Article 5 of the Paris Agreement.[31]

Brazil has a great interest in the implementation of these new instruments, because, mainly due to deforestation and fires, it is one of the top GHG emitters in the world.[32] With the implementation of *REDD+*, Brazil have been financially compensated for fighting deforestation, while at the same time contributing to the planet's climate balance.

The Ministry of Environment, the Ministry of Foreign Affairs, and the United Nations Development Program signed the project document *Payments for REDD+ results achieved by Brazil in the Amazon biome*. The project was presented by the Ministry of Environment during the first meeting of the National Council of the Legal Amazon, which endorsed the proposal and recognized the important contribution of the Green Climate Fund for immediate actions by the Federal Government to preserve and recover the Amazon Forest. This contribution has made possible the Pilot Program of Incentives to Environmental Services for the Conservation and Recovery of Native Vegetation, created by the Ministry of Environment and entitled *Floresta+*.

---

[29] Un-Redd Programme (2024b).

[30] Un-Redd Programme (2024a).

[31] Un-Redd Programme (2024a).

[32] China, the United States, India, the EU27, Russia and Brazil were the world's largest GHG emitters in 2023. Together they account for 49.8% of global population, 63.2% of global gross domestic product, 64.2% of global fossil fuel consumption and 62.7% of global GHG emissions. Among these top emitters, in 2023 China, India, Russia and Brazil increased their emissions compared to 2022. Brazil accounts for approximately 2.45% of total global emissions (European Commission 2024).

The project aims to encourage the conservation of areas additional to those provided for in the Brazilian Forest Protection Code and the recovery of small rural properties, as well as in the collective territories of Indigenous peoples and traditional communities, besides supporting innovative actions and measures to develop, implement and leverage public policies and an economy of conservation and recovery of native vegetation. Another positive result is the strengthening of the National Strategy for *REDD+* through improvements in its governance, structure, and systems.[33]

After this brief analysis of the international law on climate change, let us see how Brazilian law deals with the subject.

13.4 Brazilian Law on Climate Change

Brazil has taken part in all major international agreements on climate change: from the Framework Convention on Climate Change (UNFCCC) of 1992 and the Kyoto Protocol in 1997, to the Paris Agreement in 2015 and subsequent Conferences.

The Brazilian Constitution does not yet include a specific provision on climate change, as it was promulgated in 1988. However, a proposed constitutional amendment[34] seeks to expressly incorporate the climate agenda into the constitutional text. Regardless, prominent legal scholars have argued that, based on the current Brazilian and international legal frameworks, it is already possible to recognize the existence of an autonomous fundamental (and constitutional) right to a clean, healthy, and safe climate.[35] This doctrinal recognition, as will be observed below, has influenced significant judicial decisions in climate litigation.

In 2009, Brazil made a voluntary climate commitment to reduce GHG emissions to between 36.1% and 38.9% for the period until 2020. Although the document was a mere political declaration, with no binding nature, the announced goal was affirmed under Art. 12 of Law n. 12.187/2009, which established the National Policy on Climate Change.

This provision was reiterated in Art. 19, § 1, I, of Decree n. 9.578/2018, committing Brazil to limit its annual deforestation rate to a maximum of 3.925 km^2 by 2020. Notably, in Brazil, land-use change and deforestation are among the primary sources of greenhouse gas (GHG) emissions. Upon ratifying and incorporating the Paris Agreement, Brazil committed to reducing its GHG emissions by 37% by 2025, and by 43% by 2030, compared to 2005 levels.

However, the country has, in fact, been moving in the opposite direction of its commitments and the mitigation of climate change. Starting in 2013, annual rates of deforestation, the main driver of Brazil's greenhouse gas emissions, have steadily

[33] More about this project: PNUD Brasil (2020).

[34] Proposed Constitutional Amendment (*Proposta de Emenda Constitucional—PEC*) n. 31/2024.

[35] Fensterseifer et al. (2023), p. 148.

climbed again. Furthermore, in 2022, greenhouse gas emissions in Brazil reached a 19-year high, with deforestation being the major driving force behind the rise in emissions.[36] Since the signing of the Paris Agreement in 2015, Brazil has been increasing its greenhouse gas emissions instead of reducing them to meet the proposed target, primarily due to forest fires.

In the same way, Brazil provides huge subsidies for fossil fuels. Indeed, Brazil spent R$118.2 billion on fossil fuel subsidies in 2021, and this figure may be even higher in 2022 when considering the tax exemptions granted at the end of 2022.[37]

The Brazilian economy, therefore, is still based on fossil fuels and there is little budgetary and scientific investment for renewable sources of energy. The participation of wind and solar energy is still very small when compared to developed countries, despite the great opportunities that exist in this field. It is still up to the Brazilian state to turn this game around and move towards sustainable, low-carbon development.[38]

Even before the Paris Agreement, the National Policy on Climate Change (Law n. 12.195/09) came into effect in Brazil. According to Art. 3, the National Policy on Climate Change and the actions resulting from it will be carried out under the responsibility of federative entities and public administration bodies, observing the principles of precaution, prevention, citizen participation, sustainable development, and *common but differentiated responsibilities*, the latter in the international sphere. The National Policy on Climate Change establishes that the guidelines to be observed are, among others, the commitments assumed by the country in the United Nations Framework Convention on Climate Change, the Kyoto Protocol, and other documents on climate change (Law n. 12.195/09, Art. 5, I).

Law n. 12.195/09 is full of abstractions, principles, objectives, and concepts. A primary criticism of the statute is that it establishes no specific goals or obligations for the private sector, instead offering only general guidelines for the government to follow. Even these government-focused rules are vague and abstract, making judicial scrutiny of administrative actions and inactions more challenging. Still, the statute marked a significant step forward as a milestone in addressing climate change and global warming.[39] Importantly, it acknowledges anthropogenic climate change, defining it as a phenomenon directly or indirectly attributable to human activity that alters the composition of the global atmosphere, in addition to changes caused by natural climate variability over comparable periods (Art. 2, VIII).

The statute also reflects concern for the social and economic impacts of greenhouse gas reduction efforts. It establishes that measures to combat global warming should take into account the varying socio-economic contexts of their application, distribute resulting burdens and costs equitably among economic sectors, populations, and communities, and balance individual responsibilities regarding

---

[36] SEEG (2024).

[37] According to a study released in: Instituto De Estudos Socioeconômicos - INESC (2023).

[38] Wedy (2023).

[39] Wedy (2023).

the sources of emissions and their climate impacts (Art. 3, III). The National Policy on Climate Change is tasked with aligning economic and social development with climate protection (Art. 4, I) and ensuring that its objectives are compatible with sustainable development, aiming for economic growth, poverty eradication, and the reduction of social inequalities (Art. 4, sole paragraph). In a country with such profound social inequality, it is natural that the climate policy takes social and economic contexts into account. In other words, advancing *climate justice* is crucial.

Given the key role of deforestation in Brazil's emissions, Law n. 12.195/09 has paid special attention to the role of *carbon sinks*. Thus, the statute defines a *carbon sink* as "a process, activity or mechanism that removes greenhouse gas, aerosol or precursor of greenhouse gas from the atmosphere" (Art. 2, IX). The National Policy on Climate Change also includes objectives such as strengthening greenhouse gas removal by natural sinks within the national territory (Art. 4, IV), consolidating and expanding legally protected areas, and promoting reforestation and vegetation recovery in degraded areas (Art. 4, VII).

The statute also aims to set conditions for the reduction of greenhouse gas emissions. To this end, the statute describes the term *mitigation* as technological changes and replacements that reduce the use of resources and emissions per unit of production, as well as the implementation of measures that reduce greenhouse gas emissions and increase sinks (Art. 2, VII). Climate policy should aim to reduce anthropogenic greenhouse gas emissions across different sources (Art. 4, II). A decree from the Executive is to establish sectoral plans for climate change mitigation and adaptation, in alignment with the National Policy on Climate Change. These plans are intended to support the development of a low-carbon economy by setting quantifiable and verifiable reduction targets for anthropogenic emissions, considering the unique characteristics of each sector. These targets can be achieved through mechanisms such as the *Clean Development Mechanism* and *Nationally Appropriate Mitigation Actions* (Art. 11, sole paragraph).

The legislation was developed, first, under Decree n. 7.390/2010, later replaced by Decree n. 9.578/18, which has consolidated regulations promulgated by the Federal Executive Branch. Such regulations provide for the National Fund on Climate Change and the National Policy on Climate Change.

The legislator aimed to expand the use of economic instruments to encourage sustainable and less polluting practices. Accordingly, a guiding principle was set forth to employ financial and economic tools to support climate change mitigation and adaptation efforts (Art. 5, VII). The statute outlines several economic mechanisms, including: (a) the adoption of fiscal and tax measures to stimulate emissions reduction and greenhouse gases removal, including differentiated tax rates, exemptions, compensation, and incentives; (b) credit and financing lines from public and private financial institutions; (c) financial and economic mechanisms for climate change mitigation and adaptation within the scope of the United Nations Framework Convention on Climate Change, the Kyoto Protocol, and on a national level (Art. 6, VI, VII, X, and XI).

Law n. 12.195/09 encourages official financial institutions to offer specific credit and financing lines to support actions and activities that fulfill the objectives of the

National Policy on Climate Change. These provisions aim to induce private agents to comply with and implement the legislation, aligning with their operational scope and social responsibilities (Art. 8).

On environmental taxation, it is appropriate to recall the provision of Art. 225, § 1, VIII of the Constitution, which stipulates that the government has a duty to maintain a favorable fiscal regime for biofuels and low-carbon hydrogen, as established by Supplementary Law, to ensure that their taxation remains lower than that imposed on fossil fuels, thus securing a competitive differential in their favor. This constitutional rule has not yet been implemented.

Another important tool introduced by Law n. 12.187/09 is the evaluation of environmental impacts on both microclimates and macroclimates (Art. 6, XVIII). This assessment, ideally conducted before the issuance of environmental licenses, still requires more precise regulation. This means that the contribution of each project to global warming should be measured in advance. In other words, the indirect impacts and cumulative effects of economic activities must be considered before environmental licenses are granted.

In response to the devastating floods that struck Rio Grande do Sul in May 2024, the National Congress enacted Law n. 14,904/24, which sets up guidelines for the development of climate change adaptation plans. This legislation aims to implement measures to reduce the vulnerability and risk exposure of environmental, social, economic, and infrastructure systems to both current and anticipated adverse effects of climate change.

It is also pertinent to address the emerging and promising carbon market in Brazil, which will be analyzed in the next section.

13.5 Brazilian Greenhouse Gas Emissions Trading System

Brazil has yet to have a fully functioning regulated carbon market in place. But there have been some advances. The pressure exercised by Brazilian society, on one hand, and the Supreme Court, on the other, for prohibiting contingencies in the National Fund on Climate Change (Climate Fund) has broadened the debate and certainly resulted in regulations aimed at adopting mitigation measures, such as the structuring of the carbon market. Indeed, the Climate Fund was originally created by the Law n. 12.114/2009 and regulated by the Decree n. 9.578/2018 to define the application of its resources, including activities of GHG emissions reduction projects, and forms of technology dissemination.[40]

Law n. 12.187/09 brought a generic statement about the carbon market, prescribing that the *Brazilian Market for Emissions Reduction* (*Mercado Brasileiro de Redução de Emissões*—MBRE) will be operationalized in commodities and futures exchanges, stock exchanges, and organized over-the-counter entities, authorized by

[40] Amaral et al. (2022).

the Brazilian Securities and Exchange Commission (*Comissão de Valores Mobiliários*—CVM[41]), where the negotiation of securities representing certified reduced greenhouse gas emissions will take place (Art. 9).

By means of Decree n. 11,075/22, the Federal Government had established the *National System for the Reduction of Greenhouse Gas Emissions* and aimed to regulate the procedures for developing plans and strengthening systems to operationalize the carbon market in the country. Nevertheless, the Government reversed this decision, repealing Decree n. 11.075/22 with Decree n. 11.550/23. The repealed Decree was found to lack support from more detailed legislation, suffer from structural issues, be overly general, and potentially create legal uncertainty.

Currently, although the voluntary carbon market operates in Brazil, the regulation of the mandatory carbon market awaits the Presidential approval of a bill recently passed by the National Congress.[42]

The bill aims to introduce a regulated market and a voluntary market for securities representing the emission or removal of greenhouse gases. The proposal provides for the establishment of emission limits for different economic sectors, requiring companies to pay for pollution exceeding the permitted levels by purchasing carbon credits.

The approval of the bill by the National Congress comes shortly after the adoption of certain regulations at COP29 to facilitate an international carbon market managed by the United Nations, following years of deadlock. These regulations pertain solely to the principles and basic criteria for the methodology of accounting for carbon removal between countries. Due to the many disagreements still existing among nations, a series of issues was deferred for deliberation at future conferences.

The regulated carbon market will be gradually implemented over 6 years. Referred to as the Brazilian Greenhouse Gas Emissions Trading System (*Sistema Brasileiro de Comércio de Emissões de Gases de Efeito Estufa—SBCE*), this market will enable the trading of Brazilian Emission Quotas (*Cotas Brasileiras de Emissão—CBE*) and Verified Emission Reduction or Removal Certificates (*Certificados de Redução ou Remoção Verificada de Emissões—CRVE*).

According to the bill, each quota or *CRVE* will represent one ton of carbon dioxide equivalent (tCO2e). In the case of the regulated market, at the end of commitment periods, companies with regulated activities will be required to assess their net emissions (total gases emitted *minus* reductions or captures), and a cancellation process will occur for both types of credits (a reduction or removal certificate allows for the cancellation of an emission quota).

The concept is that, following a period of adaptation, economic activities facing greater challenges in reducing emissions through technological means will purchase

[41] Brazilian Securities and Exchange Commission or "*Comissão de Valores Mobiliários*" (CVM) is an agency under a special regime (similar to the U.S. Securities and Exchange Commission), linked to the Ministry of Economy, created in 1976 by Law n. 6.385/76, with the purpose of inspecting, regulating, and developing the stock market in Brazil.

[42] Bill (*Projeto de Lei*) n. 182/2024.

pollution quotas and certificates verifying the capture of emitted greenhouse gases, thereby achieving net-zero emissions. When buying and selling occur within financial and capital markets, they will be subject to regulation by the Brazilian Securities and Exchange Commission (*Comissão de Valores Mobiliários—CVM*). However, private placements of these assets may also take place (voluntary market).

Activities emitting over 10,000 tCO2e annually will be subject to oversight, with obligations varying by emission level. Two categories of companies are eligible to participate in the SBCE: (a) *Companies emitting between 10,000 and 25,000 tCO2e annually*: These companies will not be assigned reduction targets but must report their emissions and develop an emissions reduction plan; (b) *Companies emitting more than 25,000 tCO2e annually*: These companies must fulfill the same reporting and planning requirements but will also face mandatory emission reduction obligations.

These emission thresholds may be adjusted upward, taking into account the cost-effectiveness of the regulation and Brazil's commitments under the United Nations Framework Convention on Climate Change. However, these obligations will only apply to activities with established methodologies for measuring and verifying emissions, as defined by the governing body of the SBCE.

The mechanism established by the bill requires regulated activities to meet emission reduction targets within a specified commitment period. To detail what has been emitted, a monitoring plan will serve as a document outlining the system for measuring and verifying emissions. For entities with high emissions that need to offset them through certificates of capture or reduction of greenhouse gases, a periodic reconciliation of obligations will be required at the end of the commitment period. At this point, they must hold sufficient certificates to balance their emissions with reductions or removals of greenhouse gases.

The so-called National Allocation Plans (*Planos Nacionais de Alocação*) must establish gradual targets and the trajectory of emission limits for each commitment period defined by the law. For each period, a new plan must specify the total volume of Brazilian Emission Quotas (CBEs) and the maximum percentage of Verified Emission Reduction or Removal Certificates (CRVE) permitted in the market.

A controversial aspect of the proposed law lies in the exclusion of the agricultural sector from regulation, despite its contribution of almost 30% to the country's total emissions in 2020. Additionally, indirect emissions of carbon dioxide and other greenhouse gases resulting from the production of agricultural inputs (such as fertilizers) or raw materials will not be considered for imposing obligations to reduce greenhouse gas emissions.

Commentators argue that this exclusion hinders structural reforms, fosters inequality, and compromises the credibility of the system. The primary reason for this exclusion lies in the substantial technical challenges involved in measuring emissions within the sector. Consequently, the agribusiness sector has similarly been excluded from other regulated markets worldwide.[43]

---

[43] Consultor Jurídico (2024) and Instituto Socioambiental (2024).

Another criticism of the bill lies in the fact that representative entities of Indigenous, traditional, and *Quilombola* peoples were not consulted during its drafting, which contravenes the Indigenous and Tribal Peoples Convention of International Labour Organization (ILO) n. 169. It is also argued that the minimum allocation of funds to traditional communities for projects conducted on their lands is significantly inadequate.[44]

It is crucial to focus on Indigenous and traditional communities in the context of the carbon market, as many have been victims of questionable or fraudulent projects. Due to their well-preserved state, the territories of these groups are particularly well-suited for initiatives designed to generate forest carbon credits.[45]

Despite its criticisms, the bill introduces significant advancements. Carbon credit projects on Indigenous and traditional territories will require the free, prior, and informed consent of the affected populations, as mandated by ILO Convention n. 169. These consultations must adhere to applicable protocols or plans where they exist, with costs borne by the interested companies. Oversight and participation will be ensured by relevant official bodies, including the Ministry of Indigenous Peoples, the National Foundation of Indigenous Peoples (*Fundação Nacional do Índio*—FUNAI), and the Federal Public Prosecutor's Office. Ownership of carbon credits will be vested in those holding usufruct rights to the land, ensuring that this prerogative is principally retained by the communities.

The bill also guarantees that communities will receive and participate in managing financial resources generated from carbon credit projects conducted on their territories. They are entitled to 50% of the credits for conventional greenhouse gas removal initiatives and 70% of the credits for projects involving avoided deforestation, forest degradation, conservation, and forest management (REDD+). Additionally, the proposed law ensures compensation for Indigenous and traditional communities for collective, material, and immaterial damages resulting from carbon credit generation projects and programs.

In addition to this national and international regulatory framework, climate litigation and court rulings in climate disputes are increasingly proliferating.

13.6 Climate Litigation in Brazil

The outbreak of disputes arising from the adverse effects derived from global warming, and the need to entrust the Judiciary with the resolution of these disputes, have led to the dissemination of climate change litigation in many countries, with different settings.

UNEP, in a renewed partnership with Columbia Law School's Sabin Center for Climate Change Law, recently released an important report on the state of global

[44] Consultor Jurídico (2024).

[45] Instituto Socioambiental (2024).

climate litigation called *Global Climate Change Litigation Report—2020 Status Review*, which updates upon the last and pioneering 2017 report, *The Status of Climate Change Litigation—A Global Review*. As can be read from the Report, "Climate ambition in countries around the world remains inadequate to meet the challenge of climate change. As a result, individuals, communities, nongovernmental organizations (NGOs), business entities, and subnational governments have turned to the courts to seek relief through the enforcement of existing climate laws; integration of climate action into existing environmental, energy, and natural resources laws; clear definitions of fundamental climate rights and obligations; and compensation for climate harms. As these actions become more frequent in their occurrence, and more numerous overall, the body of legal precedent grows, forming an increasingly coherent field of law."[46]

This report also describes five types of climate cases that suggest where global climate change litigation may be heading in the coming years. *First*, plaintiffs are increasingly filing consumer and investor fraud claims alleging that companies failed to disclose information about climate risk or have misleadingly disclosed information. *Second*, recent years suggest a growing number of pre and post-disaster cases premised on a defendant's failure to properly plan for or manage the consequences of extreme weather events. *Third*, as more cases are filed and some get to an outcome, implementation of courts' orders will raise new challenges. *Fourth*, courts and litigants increasingly will be summoned to address the law and science of climate attribution as cases seeking to assign responsibility for private actors' contributions to climate change and cases arguing for greater government action to mitigate GHG emissions. *Finally*, litigants are increasingly bringing claims before international adjudicatory bodies, which may lack enforcement authority but whose declarations can shift and inform judicial understanding.[47]

Daebel and Kahl, however, suggest a simpler classification, between two broad categories of climate litigation: (a) the first category includes claims against the state, which aim to force the state to introduce more protective climate protection laws; (b) in cases under the second category, private parties take legal action against other private parties or the state because they demand compensation for damages caused by climate change.[48]

Pioneering experience on the judicial recognition of the urgency and seriousness of climate change and the state's duty to implement measures to control emissions is represented by the case *Massachusetts v. EPA*, judged by the US Supreme Court in 2007.[49] The judgment, therefore, marked an important paradigm in comparative law in which a court, besides admitting the existence and emergence of climate change according to the scientific consensus reached by the IPCC, recognized the

[46] United Nations Environment Programme (2020).
[47] United Nations Environment Programme (2020).
[48] Daebel and Kahl (2019).
[49] Justitia US Supreme Court (2007).

Environmental Regulatory Agency's positive obligation to regulate harmful emissions from new motor vehicles.

The development of climate litigation has continued to progress, with increasingly promising decisions emerging in comparative law.

On December 20, 2019, in a truly historic outcome, the Dutch Supreme Court upheld the previous decisions in the *Urgenda Climate Case*, finding that the Dutch Government has obligations to reduce emissions urgently and significantly in line with its human rights obligations.[50] Another case worth mentioning is the Irish Supreme Court's decision in *Friends of the Irish Environment v. Ireland*, handed down on July 31, 2020, which ruled that the *National Mitigation Plan in 2017* violated Ireland's Climate Action and Low Carbon Development Act 2015, the Constitution of Ireland, and obligations under the European Convention on Human Rights, particularly the right to life and the right to private and family life. The Court determined that the Plan falls short of the sort of specificity that the Act requires because a reasonable reader of the Plan would not understand how Ireland will achieve its 2050 goals.[51]

More recently, we can mention the case *Neubauer, et al. v. Germany*, decided in 2021 by the German Constitutional Court (*Bundesverfassungsgericht*). In its judgment, the Court stroked down parts of Germany's Federal Climate Protection Act (*Bundesklimaschutzgesetz*—"KSG") as incompatible with fundamental rights for failing to set sufficient provisions for emission cuts beyond 2030. The Court found that Art. 20a of the Basic Law not only obliges the legislature to protect the climate and aim towards achieving climate neutrality, but "also concerns how environmental burdens are spread out between different generations." For the first time in its jurisprudence, the Court stated that "the fundamental rights – as intertemporal guarantees of freedom – afford protection against the greenhouse gas reduction burdens imposed under Art. 20a of the Basic Law being unilaterally offloaded onto the future." It further stated that the KSG's emission provisions in question constituted an "advance interference-like effect," which possibly violates the complainants' fundamental rights and thus renders the complaints admissible.[52]

In contrast, in the landmark case of *Juliana v. United States*, climate advocacy did not prevail. In this lawsuit, a group of young plaintiffs argued that the U.-S. government's actions contributing to climate change violated their constitutional rights to life, liberty, and property. They sought a court order requiring the government to develop a comprehensive plan to reduce carbon dioxide emissions. However, in January 2020, the Ninth Circuit Court of Appeals dismissed the case, stating that the plaintiffs lacked standing because the courts do not have the authority to mandate such broad policy measures, which are the responsibility of the legislative and executive branches. The majority of the Court argued that "it was beyond the power of an Article III court to order, design, supervise, or implement the plaintiffs'

---

[50] Urgenda (2018).

[51] Sabin Center for Climate Change Law (2017).

[52] Sabin Center for Climate Change Law (2020) and Deutschland (2021).

requested remedial plan where any effective plan would necessarily require a host of complex policy decisions entrusted to the wisdom and discretion of the executive and legislative branches."[53]

Decisions from foreign courts, though not binding, offer valuable rationales for national judges to consider and can influence their rulings, creating an authentic "dialogue between national and international courts" to strengthen human rights protections. Such cases, along with others, address shared controversies regarding emissions regulation and climate change mitigation. Comparative law experience provides insights for these discussions and may have a persuasive effect. Nonetheless, it is essential to examine the unique possibilities and limitations of judicial intervention in environmental and climate issues within each legal and constitutional system.

As previously mentioned, the Brazilian Constitution enshrines the right of universal access to justice, meaning that any violation or threat of violation of a right can be brought before the Judiciary. This paves the way for broad judicial oversight of the actions and omissions of public authorities and private entities related to climate change. On the other hand, the lack of clear and precise rules specifying concrete obligations for public and private entities in the context of climate change may create obstacles to seeking judicial intervention.

Several lawsuits related to climate impacts are already in progress in many Brazilian courts. According to Sabin Center Climate Change Litigation Database, there are 36 cases of climate change litigation in Brazil. Such cases are scattered in state and federal courts across the country.[54]

As expected, we have many lawsuits related to the main causes of emissions in Brazil: forest burning and deforestation. In this category are found: a legal action in which the Federal Public Prosecutor's Office requests the environmental federal agency (Brazilian Institute for the Environment and Renewable Natural Resources or IBAMA) to increase command-and-control measures to reduce illegal deforestation at hotspots in the Brazilian Amazon rainforest; an action before the Supreme Court against the Federal Government for violating fundamental constitutional rights by failing to implement the national deforestation policy, and thereby contributing to dangerous climate change; lawsuits over timber exports; a lawsuit in which the Federal Government is asked to stop deforestation in a conservation unit; claims against the use of fire as a harvesting and processing method; a lawsuit against the omission of the Federal Government in fighting forest fires in the Amazon Forest and the Pantanal ecosystem; a lawsuit in which the legal issue is whether a farmer is responsible for the climate and environmental damage stemming from deforestation in the Amazon forest; and a dispute on the validity of a fine imposed by IBAMA for the burning of land for livestock grazing.

We can also mention lawsuits requesting specific actions by the government to adopt environmental and climate protection measures. This type of case includes

[53] Sabin Center for Climate Change Law (2015).

[54] Sabin Center for Climate Change Law (n.d.).

cases related to mitigation targets in the country's nationally determined contributions to reduce emissions; a claim to recognize the fundamental right to a stable climate and to compel the Federal Government to meet emission and deforestation targets; and a lawsuit asking for structural measures to protect a lagoon. Here we can also refer to the actions before the Supreme Court, challenging the non-release of resources from environmental funds or the decision-making procedures for these funds (such as the Climate Fund and the Amazon Fund). Also worth mentioning are the lawsuits (already studied above) against environmental deregulation by the National Environmental Council (CONAMA).

The exploitation of coal and thermal power plants is also the target of several lawsuits. It is also worth highlighting the legal claims filed by the government or the public prosecutor's office against individuals and companies for climate damage. All these and other climate litigation cases can be read at Columbia University's Sabin Center for Climate Change Law website, mentioned above.[55]

This context leads to an analysis of one of Brazil's key climate disputes, *ADPF 708*, concerning the *Climate Fund,* recently adjudicated by the Supreme Court.[56] This case centered on budget cuts to the National Fund on Climate Change, established under Law n. 12.114/09. The Court ruled that withholding these resources is unconstitutional, affirming that the Executive Branch has a constitutional duty to ensure the Climate Fund's operation and to allocate its resources annually for climate change mitigation. Consequently, blocking these funds is prohibited, grounded in the constitutional duty to protect the environment (Art. 225 of the Brazilian Constitution), Brazil's international rights and commitments (Art. 5, § 2), and the constitutional principle of separation of powers. The Court further underscored that environmental treaties are a form of human rights treaties and therefore carry a higher status than ordinary law (*supra-legal status*).[57] As a result, the government cannot legally choose to ignore its obligation to combat climate change.[58]

These points were restated in the Supreme Court's decision in a case involving state omission in the protection of the Legal Amazon. The Court declared the unconstitutionality of federal decrees that dissolved the Amazon Fund committees, leading to an interruption in the receipt of donations. It was held that this measure led to a retraction in the fulfillment of international commitments of a *supra-legal nature* assumed by Brazil, as well as having a direct impact on the reality of the Legal

---

[55] Sabin Center for Climate Change Law (n.d.).

[56] *ADPF* is the short of *Arguição de Descumprimento de Preceito Fundamental* or *Claim of Non-compliance with a Fundamental Precept.* This is a specific type of lawsuit, filed before the Supreme Court, whose purpose is to prevent or repair violation to a fundamental precept, resulting from an act of the government. The ADPF is provided for in the Constitution (Art. 102, § 1) and regulated by the Law n. 9.882/99.

[57] The Supreme Court has ruled (leading case: RE 466.343) that international human rights treaties have a higher value than the act of parliament (statutory law). However, it has a lower status than the Constitution.

[58] Brazilian Supreme Court, ADPF 708, Rapporteur Min. Luís Roberto Barroso, j. 1 Jul. 22.

Amazon. In this context, the change in the Amazon Fund's governance model, with the suppression of essential regulatory mechanisms for its management, represents a step backward in environmental protection.[59]

Three other landmark decisions on the protection of the Amazon Rainforest and the Pantanal, previously discussed in this book, are also related to the issue of climate change. And it could not be otherwise, as we have emphasized, since Brazilian emissions primarily stem from land-use changes, deforestation, and forest fires.

In the first case, the Supreme Court held that, at present, there is no *unconstitutional state of affairs* (or massive violations of fundamental rights) in the environmental policy for the protection of the Amazon biome, given that a process is underway to resume the effective exercise of this constitutional duty. This resumption occurred through the reactivation of the Action Plan for the Prevention and Control of Deforestation in the Legal Amazon,[60] which had been suspended by the previous administration. The Court reiterated that environmental protection is not a political option but a duty mandated by the Constitution. Although it did not declare an *unconstitutional state of affairs*, the justices acknowledged the existence of structural deficiencies in the policy for the protection of the Legal Amazon and ordered the Federal Government to undertake a *meaningful engagement* regarding the illegal deforestation of the Amazon Rainforest.[61]

In a similar case, the Supreme Court also held that there is no *unconstitutional state of affairs* in the environmental protection policy for the Amazon and Pantanal. However, it ordered certain measures to be implemented. The Federal Government was required to present, within 90 days, a publicly available plan for the prevention and control of fires in the Pantanal and Amazon. IBAMA and the state agencies were tasked with ensuring the transparency of vegetation suppression authorizations. The National Council of Justice (CNJ)[62] was directed to monitor judicial cases with a significant impact on deforestation.[63]

A final case to be mentioned concerns the allegation of omission by the National Congress regarding the enactment of a statute for the special protection of the Pantanal biome. Article 225, § 4 of the Constitution states that the "Brazilian

[59] Brazilian Supreme Court, ADO 59, Rapporteur Min. Rosa Weber, j. 3 Nov. 22.

[60] The Action Plan for the *Prevention and Control of Deforestation in the Legal Amazon* (*Plano de Ação para Prevenção e Controle do Desmatamento na Amazônia Legal* or *PPCDAm*), launched in 2004, aims to continuously reduce deforestation and establish conditions for a transition to a sustainable development model in the Legal Amazon.

[61] Brazilian Supreme Court, ADPF 760, ADO 54, j. 14 Mar. 24.

[62] In summary, the National Council of Justice (*Conselho Nacional de Justiça*—CNJ) is an administrative body within the Judiciary, aimed at improving the functioning of the Brazilian judicial system, particularly concerning administrative and procedural oversight and transparency. It does not have a jurisdictional function; that is, it does not adjudicate legal cases or appeals. The CNJ was created by Constitutional Amendment n. 45/2004, through the addition of Art. 103-B to the Constitution.

[63] Brazilian Supreme Court, ADPF 746, ADPF 857, j. 20/3/24.

Amazonian Forest, the Atlantic Forest, the Serra do Mar, the *Pantanal Mato-Grossense* and the coastal zone are part of the national patrimony, and they shall be used, as provided by law, under conditions which ensure the preservation of the environment, therein included the use of mineral resources." Therefore, the Constitution mandates the enactment of national legislation to promote the protection of these biomes, including the Pantanal. In a recent ruling on an action for unconstitutionality by omission, the Supreme Court recognized an unconstitutional omission concerning the enactment of a regulatory law for the special protection of the *Pantanal Mato-Grossense* biome. The Court set a deadline of 18 months for the Legislature to address this regulatory gap, aiming to ensure environmental preservation and the sustainable use of the natural resources of the Pantanal.[64]

The inadequacy of government actions to combat global warming, combined with the growing visibility of climate-related harms, has fueled an increase in disputes on this issue. In Brazil, the independence of the Judiciary and the Public Prosecution Service, the extensive judicial tools available to address rights violations, and comparative law examples have created fertile ground for climate litigation to be resolved by the courts. It falls to the Judiciary, alongside the Executive and Legislative branches within their respective powers, to shape climate change law.

The cases mentioned previously highlight that Brazil has become a stage for climate litigation with the potential for remarkable repercussions, giving rise to a solid scientific and, especially, constitutional debate on a topic that has gained significant importance.

References

Amaral W, Pimentel C, Wedy G (2022) The carbon market and its regulation in Brazil. Sabin Center for Climate Change Law 9-2022:1–23

Berlie AB (2018) Global warming: a review of the debates on the causes, consequences and politics of global response. Ghana J Geogr 10:144–164

Consultor Jurídico (2024) PL que cria mercado de carbono falha ao excluir agronegócio, afirmam especialistas. https://www.conjur.com.br/2024-nov-26/pl-do-mercado-de-carbono-tem-brechas-criticas-dizem-especialistas/?fbclid=IwY2xjawGz_xtleHRuA2FlbQIxMQABHez5B0qigHPlzgNEwaJott6ujJYDPR9GzjpvXuwfKyQAZKKTUEJ9My-87g_aem__2Gttdb09y3eMgUqu17PjQ. Accessed 26 Nov 2024

Conway EM, Oreskes N (2011) Merchants of doubt: how a handful of scientists obscured the truth on issues from tobacco smoke to global warming. Bloomsbury Press, New York

Curry JA (2019) Climate change: what we know, what we don't know. Guerra e Paz, Lisboa

Daebel M-C, Kahl W (2019) Climate change litigation in Germany: an overview of politics, legislation and especially jurisdiction regarding climate protection and climate damages. Eur Energy Environ Law Rev 28(2):67–76

Deutschland (2021) Bundesverfassungsgericht. Verfassungsbeschwerden gegen das Klimaschutzgesetz teilweise erfolgreich. https://www.bundesverfassungsgericht.de/SharedDocs/Pressemitteilungen/DE/2021/bvg21-031.html. Accessed 18 Nov 2024

[64] Brazilian Supreme Court, ADO 63/MS, j. 6 Jun. 24.

European Commission (2024) *EDGAR* - Emissions Database for Global Atmospheric Research, 2024 Report. https://edgar.jrc.ec.europa.eu/report_2024. Accessed 18 Nov 2024

Fensterseifer T, Sarlet IW, Wedy G (2023) Curso de Direito Climático. Revista dos Tribunais, São Paulo

Instituto De Estudos Socioeconômicos - INESC (2023) Estudo lançado na COP 27 revela que Brasil perde quase R$ 120 bi para subsidiar combustíveis fósseis. https://www.inesc.org.br/estudo-lancado-na-cop-27-revela-que-brasil-perde-quase-r-120-bi-para-subsidiar-combustiveis-fosseis/. Accessed 18 Nov 2024

Instituto Socioambiental (2024) Câmara aprova e vai à sanção presidencial projeto que cria mercado de créditos de carbono. https://www.socioambiental.org/noticias-socioambientais/camara-aprova-e-vai-sancao-presidencial-projeto-que-cria-mercado-de. Accessed 20 Nov 2024

Intergovernmental Panel on Climate Change (IPCC) (1990) Policymakers summary. https://www.ipcc.ch/site/assets/uploads/2018/03/ipcc_far_wg_I_spm.pdf. Accessed 18 Nov 2024

Intergovernmental Panel on Climate Change (IPCC) (2021) Summary for policymakers. https://www.ipcc.ch/report/ar6/wg2/downloads/report/IPCC_AR6_WGII_SummaryForPolicymakers.pdf. Accessed 18 Nov 2024

Justitia US Supreme Court (2007) Massachusetts v. EPA, 549 U.S. 497. https://supreme.justia.com/cases/federal/us/549/497/#tab-opinion-1962180. Accessed 18 Nov 2024

Leape JP, Miller AS, Percival RV, Schroeder CH (2018) Environmental law: law, science, and policy. Wolters Kluwer, New York

Leite GB, Fernandes EA (2021) Atuação dos projetos de mecanismo de desenvolvimento limpo para o desenvolvimento sustentável no Brasil. Revista de Economia Política 41(2):351–371

Oreskes N (2019) Why trust science? Princeton University Press, Princeton

PNUD Brasil (2020) Projeto "Pagamentos por resultados de REDD+ por resultados alcançados pelo Brasil no bioma Amazônia." https://www.undp.org/pt/brazil/news/projeto-%E2%80%9Cpagamentos-por-resultados-de-redd-por-resultados-alcan%C3%A7ados-pelo-brasil-no-bioma-amaz%C3%B4nia%E2%80%9D. Accessed 18 Nov 2024

Ribeiro WC, de Souza GD (2009) NovaGerar: experiência pioneira do Brasil no MDL. Cronos 10(2):15–34

Sabin Center for Climate Change Law (2015) Climate change litigation databases. Juliana v. United States. http://climatecasechart.com/case/juliana-v-united-states/?cnreloaded=1. Accessed 18 Nov 2024

Sabin Center for Climate Change Law (2017) Climate change litigation databases. Friends of the Irish Environment v. Ireland. http://climatecasechart.com/non-us-case/friends-of-the-irish-environment-v-ireland/. Accessed 18 Nov 2024

Sabin Center for Climate Change Law (2020) Climate change litigation databases. Neubauer, et al. v. Germany. http://climatecasechart.com/non-us-case/neubauer-et-al-v-germany/. Accessed 18 Nov 2024

Sabin Center for Climate Change Law (n.d.). https://climatecasechart.com/non-us-jurisdiction/brazil/. Accessed 18 Nov 2024

SEEG (2024). https://plataforma.seeg.eco.br/. Accessed 18 Nov 2024

United Nations Climate Change (2021) Federative Republic Of Brazil. Paris Agreement Nationally Determined Contribution (NDC), 21 March 2022. https://unfccc.int/sites/default/files/NDC/2022-06/Updated%20-%20First%20NDC%20-%20%20FINAL%20-%20PDF.pdf. Accessed 18 Nov 2024

United Nations Climate Change (2022) Project: 0008 Brazil NovaGerar landfill gas to energy project - crediting period renewal request. https://cdm.unfccc.int/Projects/DB/DNV-CUK1095236970.6/view. Accessed 18 Nov 2024

United Nations Climate Change (2024a) A brief history of the climate change process. https://unfccc.int/cop7/issues/briefhistory.html. Accessed 18 Nov 2024

United Nations Climate Change (2024b) Emissions trading. https://unfccc.int/process/the-kyoto-protocol/mechanisms/emissions-trading. Accessed 18 Nov 2024

United Nations Climate Change (2024c) History of the convention. https://unfccc.int/process/the-convention/history-of-the-convention#Essential-background. Accessed 18 Nov 2024

United Nations Climate Change (2024d) Joint implementation. https://unfccc.int/process/the-kyoto-protocol/mechanisms/joint-implementation. Accessed 18 Nov 2024

United Nations Climate Change (2024e) Nationally Determined Contributions (NDCs). https://unfccc.int/ndc-information/nationally-determined-contributions-ndcs#NDC-registry--. Accessed 18 Nov 2024

United Nations Climate Change (2024f) The clean development mechanism. https://unfccc.int/process-and-meetings/the-kyoto-protocol/mechanisms-under-the-kyoto-protocol/the-clean-development-mechanism. Accessed 18 Nov 2024

United Nations Climate Change (2024g) The Paris Agreement. https://unfccc.int/process-and-meetings/the-paris-agreement/the-paris-agreement. Accessed 18 Nov 2024

United Nations Climate Change (2024h) What is the Kyoto Protocol? https://unfccc.int/kyoto_protocol. Accessed 18 Nov 2024

United Nations Environment Programme (2020) Global Climate Litigation Report: 2020 status review. https://wedocs.unep.org/bitstream/handle/20.500.11822/34818/GCLR.pdf?sequence=1&isAllowed=y. Accessed 18 Nov 2024

Un-Redd Programme (2024a) About REDD+. https://www.un-redd.org/about/about-redd. Accessed 18 Nov 2024

Un-Redd Programme (2024b) The programme. https://www.un-redd.org/about/programme. Accessed 18 Nov 2024

Urgenda (2018) Landmark decision by Dutch Supreme Court. https://www.urgenda.nl/en/themas/climate-case/

Wedy G (2023) Litígios Climáticos, 2nd edn. Forum, Belo Horizonte

World Bank (2004) Brazil - Nova Gerar Landfill Rio de Janeiro. https://documents1.worldbank.org/curated/en/423711468770141049/pdf/PID0P079182.pdf. Accessed 18 Nov 2024

Yergin D (2012) The Quest: energy, security, and the remaking of the modern world. Penguin Publishing Group, New York

Chapter 14
National Water Resources Policy

14.1 Introduction

Water, a natural resource of utmost importance for human survival, is essential for vital functions and exists in the biosphere in liquid (fresh and salt), solid (fresh), and vapor (fresh) form. Its liquid form constitutes about 97.72% of the water in the biosphere, of which 97% is salty and only 0.72% is fresh.[1]

Water is essential for life as well as for the promotion of sustainable development, especially food production in agriculture. Humanity will need to manage water resources sustainably to provide a dignified existence for present and future generations. The issue of water is one of the most discussed and relevant topics in the contemporary world, with scholars predicting that it may become the next major trigger for crises and conflicts globally.

Brazil plays a critical role in the global water context, being one of the countries with the highest availability of water resources. It holds 12% of the planet's freshwater reserves and accounts for 53% of South America's water resources. Additionally, transboundary river basins cover 60% of Brazilian territory. However, Brazil has seen a 15.7% reduction in its surface water area since the early 1990s, shrinking from nearly 20 million hectares to 16.6 million hectares by 2020. The country is also home to three of the world's largest aquifers: the *Amazon Basin*, the *Guarani Aquifer System*, and the *Maranhão Basin*.

[1] Fiorillo (2017), p. 282.

R. Martins Costa Moreira, G. Wedy, *Brazilian Environmental and Climate Change Law*, https://doi.org/10.1007/978-3-031-87998-2_14

14.2 Constitutional and Legal Protection of Water Resources

The first Brazilian legislation addressing water resources was the *Water Act*[2] (Decree n. 24.643/34). At the time, this legislation primarily focused on protecting private interests and economic rights, as shown by its emphasis on the *commercial* use of water. The Decree paid particular attention to industrial water usage and, above all, to hydraulic energy.

The *Water Act* classified waters into public (common or dominical use), common, and private. The Decree also provided for groundwater, but, subsequently, the Mining Act (Decree n. 227/1967) considered it as a mining deposit, regulated by a specialized statute. It should be noted, however, that groundwater, by the new legal framework, was included in the water resources management system.

This classification, however, is no longer compatible with the current Constitution, by which all waters are considered to belong to the public domain, including underground waters, and there are no longer common or private waters. Besides, the waters are under the domain of the federal, state, and federal district governments. Indeed, there have been no longer municipal waters since the 1946 Constitution. In fact, the *publicization of water* finds its source not only under the 1988 Constitution (Art. 20, III and Art. 26), but also under Law n. 9.433/97 (Art. 1, I), which establishes the National Water Resources Policy. In general, the former Water Act has been largely superseded and can only be applied in a supplementary manner and in limited cases.

The Superior Court of Justice has embraced this approach, deciding that the compulsory purchase of a property located on a riverbank does not warrant compensation, as rivers and their banks are in the public domain. The Court held that the old Water Act (Decree n. 24.643/1934) must be interpreted in light of the 1988 Constitution and Law n. 9433/1997 (Water Resources Policy Act), which recognizes only public ownership of water resources. Pursuant to Art. 20, III, and Art. 26, I, of the Constitution, private ownership of lakes, rivers, surface, or underground waters—whether flowing, emerging, or in storage—as well as any waterways, has been abolished. Article 31 of the 1934 Water Act is thus interpreted to imply state ownership over all banks of state rivers, as specified in Art. 26 of the Constitution, with federal ownership over federal rivers (Art. 20), given that neither municipal nor private rivers exist anymore.[3]

With technological advancements, increased human intervention in the environment and water systems, ecological crises, and a heightened focus on environmental balance, the need to protect not only water quantity but also quality has become evident. In line with this, Law n. 6.938/81, in Art. 8, VII, assigned the National Environmental Council (CONAMA) the authority to set rules, criteria, and standards

---

[2]Called in Brazil as *Water Code* (*Código de Águas*).

[3]Superior Court of Justice, REsp 1352673/SP, Rapporteur Min. Herman Benjamin, j. 06 Mai 2014.

for environmental quality control and maintenance, aimed particularly at the rational use of environmental resources, including water. Subsequently, CONAMA issued Resolution n. 20/1986, focusing on water quality management. This Resolution was later replaced in 2005 by Resolution n. 357, which classifies bodies of water based on salinity into freshwater, saline, and brackish categories for quality management purposes. Additionally, Law n. 9.984/00 established the National Water Agency (ANA), a federal entity tasked with implementing the National Policy for Water Resources and coordinating the National System for Water Resources Management.

The Brazilian Constitution included as property of the Federal Government the lakes, the rivers, and any waterways in lands under its domain, or that border more than one state, serve as limits with other countries, or extend to or from foreign territory, as well as marginal lands and river beaches; the river and lake islands bordering other countries; the maritime beaches; the oceanic and coastal islands, excluding from these the areas referred to under Art. 26, II; the natural resources of the continental shelf and the exclusive economic zone; the territorial sea; and the potential for hydraulic energy (Art. 20).

According to Art. 26 of the Constitution, the state's property includes: I—surface or subterranean waters, flowing, emerging or in deposit, with the exception, in this case, of those resulting from work carried out by the Federal Government, as provided by law; II—the areas, on ocean and coastal islands, which are within their domain, excluding those under the domain of the Federal Government, the municipalities or third parties; III—the river and lake islands which do not belong to the Federal Government. Therefore, water resources generally fall under state jurisdiction, except for those waters designated to the Federal Government by Art. 20 of the Constitution.

14.3 Administrative and Legislative Competence

In this chapter, we will analyze issues related to administrative and legislative competence regarding the use and protection of water resources.

It is within Federal Government's authority, as established under Art. 21 of the Constitution, the operation, directly or through authorization, concession or permission (XII), of electric power services and facilities and the energetic exploitation of watercourses, jointly with the states wherein those hydro energetic potentials are located ("b"); railway and waterway services between seaports and national borders or which cross the boundary of a state or territory ("d"); and sea, river and lake ports; ("f").

The National Congress is also responsible for establishing the national system for water resource management and setting criteria for granting rights to their use (Art. 21, XIX). Moreover, the National Congress holds exclusive authority to legislate on water resources (Art. 22, IV). However, all the federative entities have the authority to register, monitor, and inspect the granting of rights to research and exploit water and mineral resources in their territories (Art. 23, XI).

Despite the National Congress's exclusive authority to legislate on water resources, the Supreme Court has ruled that states, alongside the Federal Government, are also empowered to legislate on environmental protection (Constitution, Art. 24, VI and VIII), which naturally includes the protection of water resources. Thus, states may regulate water use charges within their territories.[4]

Conversely, states cannot exempt agricultural, agro-industrial, and rural water use from charges, nor can they waive the requirement for permits to use water resources, such as for drilling tube wells. Such provisions exceed the scope of environmental protection and infringe on the National Congress's exclusive authority to set criteria for water use permits, as outlined in Law n. 9433/97. In the same vein, these state rules contradict the constitutional duty of public authorities to regulate techniques, methods, and substances that pose risks to life, quality of life, and the environment (Constitution, Art. 225, § 1, V).[5] Likewise, states cannot block Watershed Committees from providing input to the State Water Resources Council prior to its actions, as this would limit community participation in water resource management, undermining the democratic principle (Constitution, Art. 1).[6]

The legal framework governing groundwater is particularly important given the widespread, often indiscriminate and disorganized, drilling of artesian wells. This practice has led to significant environmental harm, reduced availability of water resources for current and future generations, and public health risks due to inadequate water treatment.

The Superior Court of Justice has addressed this issue, emphasizing that constitutional provisions regarding the ownership of water resources and the authority of federative entities align with Art. 225 of the Constitution, which obliges public authorities and the community to protect and preserve the environment for present and future generations. Although groundwater is typically under the state domain (except in cases specified in Art. 20, III of the Brazilian Constitution), the state's environmental authority remains unrestricted in its legislative and regulatory powers to prevent both quantitative (overexploitation and depletion) and qualitative (aquifer contamination) degradation of water resources. This multiplicity and overlap of control are justified by the increasing scarcity of water, affecting millions of Brazilians in major cities and countless others in rural areas. This issue becomes even more concerning given dire predictions of worsening conditions and public crises in light of undeniable anthropogenic climate change.[7]

The Superior Court of Justice has also held that Art. 12, II of Law n. 9.433/97 requires government authorization for groundwater extraction, justified by its

[4]Brazilian Supreme Court, ADI 3.336, Rapporteur Min. Dias Toffoli.

[5]Brazilian Supreme Court, ADI 5025, Rapporteur Min. Dias Toffoli, j. 08 Feb. 2021; ADI 5016, Rapporteur Min. Alexandre de Moraes, j. 11 Oct. 2018.

[6]Brazilian Supreme Court, ADI 5016, Rapporteur Min. Alexandre de Moraes, j. 11 Oct. 2018.

[7]Superior Court of Justice, REsp 1306093/RJ, Rapporteur Min. Herman Benjamin, j. 28 Mai 2013.

scarcity, its status as a public domain resource, and its significant economic value.[8] Accordingly, the Court ruled that it is forbidden to extract groundwater for residential use without prior authorization and an environmental permit issued by the government.[9]

As for municipalities, although they do not hold ownership of water bodies, they are vested with the authority to oversee the exploitation of surface and groundwater resources within their territory. This power enables them to regulate and, when necessary, restrict the drilling and use of artesian wells as part of their legitimate urban, environmental, health, and consumer protection responsibilities.[10]

14.4 National Policy on Water Resources

Law n. 9.443/1997 established the National Policy on Water Resources, which has the following key features: water is a public good; water is a limited natural resource with economic value; in situations of scarcity, the priority use of water resources is for human consumption and animal feeding; the management of water resources must always provide for the multiple uses of water; the watershed is the territorial unit for the implementation of the National Policy on Water Resources and the operation of the National System on Water Resources Management; and the management of water resources shall be decentralized and rely upon the participation of the government, users, and communities (Art. 1).

The objectives of the National Water Resources Policy are to assure the current and future generations the necessary availability of water, with quality standards adequate for their respective uses; the rational and integrated use of water resources, including waterway transport, with a view to sustainable development; and the prevention and defense against critical hydrological events of natural origin or resulting from the inadequate use of natural resources (Art. 2).

The general guidelines for implementing the National Water Resources Policy include: systematic management of water resources, considering both quantity and quality; adapting water resource management to the diverse physical, biological, demographic, economic, social, and cultural conditions across regions; integrating water resource management with environmental management; coordinating water resource planning with that of user sectors, as well as with regional, state, and national planning; aligning water resource management with land use; and integrating watershed management with that of estuarine systems and coastal zones (Art. 3).

[8] Superior Court of Justice, AgRg no REsp 1.352.664/RJ, Rapporteur Min. Mauro Campbell Marques; AgRg no AgRg no REsp 1.185.670/RS, Rapporteur Min. Benedito Gonçalves.

[9] Superior Court of Justice, EREsp 1.335.535/RJ, Rapporteur Min. Herman Benjamin, j. 26 Sept. 2018.

[10] Superior Court of Justice, REsp 994.120/RS, Rapporteur Min. Herman Benjamin, j. 25 Aug. 2009.

Additionally, the Federal Government will coordinate with states to manage water resources of shared interest (Art. 4).

The tools of the National Water Resources Policy consist of: Water Resources Plans; classification of bodies of water into categories based on predominant uses; granting rights to use water resources; charging fees for water usage; and the Water Resources Information System (Art. 5).

14.5 Allocation of Water Resources

The system for granting rights to use water resources aims to ensure both quantitative and qualitative control of water exploitation and to safeguard the effective exercise of access rights to water (Art. 11).

The following uses of water resources require permission from public authorities: diversion or extraction of water from a body of water for final consumption, including public supply, or as an input in production processes; extraction of water from an underground aquifer for final consumption or as an input in production processes; discharge of sewage and other liquid or gaseous waste, treated or untreated, into a body of water for dilution, transport, or final disposal; use of hydroelectric potential; and other uses that alter the character, quantity, or quality of the water (Art. 12).

The uses below are exempt from requiring permission from public authorities, as defined by regulations: use of water resources to meet the needs of small population centers in rural areas; diversions, extractions, and discharges deemed insignificant; and accumulations of water considered to be of insignificant volume (Art. 12, § 1).

The allocation of water rights will be carried out by an act of the relevant authority within the Federal Government, the states, or the federal district. The authority to grant permissions depends on the jurisdiction over the water body. For rivers under federal jurisdiction, the National Water Agency (*Agência Nacional de Águas*, or ANA) is responsible for issuing authorizations, as specified in Art. 4, IV of Law n. 9.984/00. In the case of states and the federal district, authorization depends on their respective regulations. Furthermore, the Federal Government may delegate to the states and the federal district the authority to grant rights to use water resources under federal jurisdiction (Art. 14).

Granting of the right to use water resources may be suspended partially or totally, permanently or for a fixed term, in the following circumstances: non-compliance with the terms of the authorization; absence of use for 3 years in a row; the urgent need for water to meet disaster situations, including those resulting from adverse weather conditions; need to prevent or reverse serious environmental degradation; need to meet priority uses of collective interest for which there are no alternative sources; need to maintain the navigability conditions of the body of water (Art. 15). All water resource use rights will be authorized for a renewable period of no more than 35 years (Art. 16). The allocation does not imply partial disposal of the waters, which are inalienable, but the simple right to use them (Art. 18).

It is worth pointing out that the granting of water resource use does not exempt or replace environmental licensing. Whenever the activity is effectively or potentially polluting or capable, in any way, of causing environmental degradation, it is also subject to licensing. In addition, CONAMA's Resolution n. 01/1986 requires the preparation of an Environmental Impact Assessment for the licensing of several activities involving the use of water resources because they potentially cause significant environmental degradation. This is the case of hydraulic works for the exploitation of water resources, such as: dams for hydroelectric purposes, above 10 MW, sanitation or irrigation, the opening of canals for navigation, drainage, and irrigation, the rectification of waterways, the opening of barriers and river mouths, transposition of basins, and dams. It should be noted that, as specified in Art. 10, § 1, of CONAMA Resolution n. 237/1997, environmental licensing can only be issued if there is a prior grant of the right to use water.

14.6 Charging for the Use of Water Resources

Charging for the use of water resources is intended to acknowledge water as an economic good and provide users with an understanding of its true value; to promote the rationalization of water use; and to generate financial resources to support programs and actions outlined in water resource management plans (Art. 19). Charging for water use assumes a regulatory function, by fostering its rational and sustainable use, avoiding waste and establishing priorities.

It is appropriate to highlight a decision issued by the Superior Court of Justice, which recognized public supply as the priority use of water resources, according to Art. 1, III of Law n. 9.433/1997, a rule that is violated in the case of construction of a building in a watershed protection area.[11]

14.7 Water Resource Use Regulation

In implementing the National Water Resources Policy, the Federal Government is responsible for: taking the necessary steps to implement and operate the National Water Resources Management System; granting the right to use water resources and regulating and inspecting their use, within its sphere of competence; implementing and managing the National Water Resources Information System; and promoting the integration of water resources management with environmental management. The Federal Government shall indicate, by decree, the authority responsible for granting

[11] Superior Court of Justice, REsp 1376199/SP, Rapporteur Min. Herman Benjamin, j. 19 Aug. 2014.

the right to use water resources under the Federal Government's domain (Law n. 9.433/1997, Art. 29).

It is the duty of the state and federal district government, within their sphere of authority: to grant the rights to use water resources and to regulate and inspect their use; to carry out the technical control of hydric supply works; to implement and manage the Information System on Water Resources, at the state and federal district levels; to promote the integration of hydric resources management with environmental management (Art. 30). Finally, municipalities shall promote the integration of local policies on basic sanitation, land use, occupation and conservation, and environmental protection with federal and state policies on water resources (Art. 31).

Law n. 9.433/97 established the National Water Resources Management System with the following objectives: to coordinate the integrated management of water resources; to settle water resources disputes; to implement the National Water Resources Policy; to plan, regulate, and control the use, preservation, and recovery of water resources; to promote billing for the use of water resources (Art. 32).

The National Water Resources Management System consists of: the National Water Resources Council; the National Water Agency; the State and Federal District Water Resources Councils; the Watershed Committees; federal, state, federal district, and municipal agencies involved in water resources management; and the Water Agencies (Art. 33).

These are, in short, the main features of the regulatory and legislative framework for the protection and use of water resources in Brazil. Certainly, many conflicts on the subject will emerge in the near future.

Reference

Fiorillo CAP (2017) Curso de Direito Ambiental Brasileiro, 17th edn. Saraiva, São Paulo

Chapter 15
Conclusion

Here, we offer some concluding remarks on Brazilian environmental law. Brazilian environmental law took shape in 1980 with the establishment of the National Environmental Policy under Law n. 6.938/81. From that point forward, environmental law in Brazil evolved significantly, particularly as the 1988 Constitution imposed upon public authorities and society the duty to protect the environment for the benefit of present and future generations.

The strength of Brazilian environmental law is attributed not only to its constitutional foundations but also to other features of the Brazilian legal system, including: modern administrative tools, such as environmental licensing and Environmental Impact Assessments; various types of lawsuits and a broad array of entities with standing to bring claims before the Judiciary; and a robust system of administrative, civil, and criminal liability for environmental damage and violations. In addition to this legislative framework, the Brazilian legal system benefits from well-established legal principles shaped by literature and jurisprudence, along with an independent Judiciary and well-structured public institutions.

Yet, Brazil still has significant progress to make in areas such as the effective enforcement of environmental legislation and judicial rulings, legal certainty, climate protection, promotion of renewable energy, combating deforestation and forest fires, and protecting indigenous communities, to name a few.

As discussed in Chap. 2, the Brazilian legal system has also been shaped by developments in international environmental law. Environmental protection in Brazil has evolved gradually, leading to today's framework that not only includes a constitutional guarantee of the fundamental right to an ecologically balanced environment but also a comprehensive regulatory and statutory foundation in environmental law. However, despite this extensive and generally high-quality legislation, Brazil faces a crisis of effectiveness in enforcing environmental laws, as does much of the world. The growth of Brazilian environmental law reflects a slow legislative evolution, remarkably with the incorporation of environmental protections in the 1988 Constitution, alongside the increasing influence of the court's jurisprudence.

R. Martins Costa Moreira, G. Wedy, *Brazilian Environmental and Climate Change Law*, https://doi.org/10.1007/978-3-031-87998-2_15

In Chap. 3, we emphasized that legal principles hold significant weight in the reasoning behind judicial decisions in Brazil, as judges and courts can impose implicit duties on public authorities and private individuals derived from the Constitution. These principles serve as a foundation for establishing such duties, particularly those related to environmental protection. In practice, Brazilian lawyers, judges, and scholars now apply key principles recognized in international law, such as the sustainability principle, the precautionary principle, and the polluter-pays principle.

As mentioned in Chap. 4, Brazil has been following the same path as most countries in gradually building an environmental regulatory framework, with the 1988 Constitution marking a milestone in environmental protection. This Constitution elevated environmental protection to the level of a constitutional right, dedicating an entire chapter to the environment (Art. 225). Accordingly, the human-environment relationship is governed by Art. 225 and its paragraphs, with the *caput* of Art. 225 stating, "all have the right to an ecologically balanced environment, an asset of common use and essential to a healthy quality of life, and both the government and the community shall have the duty to defend and preserve it for present and future generations." Additionally, this chapter also discusses the legal framework governing environmental assets and the relationship between the protection of Indigenous rights and the preservation of ecosystems.

Chapter 5 addressed animal rights within the Brazilian legal system. An intrinsic value and inherent dignity of non-human life have been recognized, with many scholars even advocating for animals to be regarded as subjects of rights. In addition to a comprehensive legislative framework, court jurisprudence has increasingly embraced a more protective stance, recognizing at least an intrinsic value to animal welfare that, in some cases, may even take precedence over human interests.

In Chap. 6, we clarified that the evolution of environmental law in Brazil has sparked debates regarding the allocation of legislative and administrative authority across the three levels of the Federation. The Brazilian Constitution grants each federative entity a role in ensuring and achieving ecological protection, and distributing legislative and administrative powers in environmental matters at the federal, state, and municipal levels. While this dispersion of regulatory functions might appear to enhance environmental protection, it has led to numerous federative conflicts that often require resolution by the Supreme Court.

As outlined in Chap. 7, the National Environmental Policy (NEP) was established by Law n. 6.938/81, marking a pivotal moment in Brazilian environmental law and the legal protection of nature. This law provided a crucial framework for planning and stability in Brazilian environmental policy, giving environmental law a holistic and centralized vision, moving beyond the improvisation and fragmentation of earlier efforts. The chapter also examined the National Environmental System (*Sistema Nacional do Meio Ambiente—SISNAMA*), comprised of agencies and entities from the Federal Government, states, federal districts, and municipalities responsible for the protection and improvement of environmental quality, according to Art. 6 of Law n. 6.938/81. Special focus was given to the National Council for the Environment (*Conselho Nacional do Meio Ambiente—CONAMA*), which plays a critical role in formulating environmental regulations. This function of CONAMA raises several issues, including the limits of its regulatory authority and the composition of this governmental body.

Chapter 8 dealt with the fundamental tools used by public administration for environmental regulation. Emphasis was first placed on the traditional "command-and-control" approach, which involves administrative enforcement of environmental laws and the imposition of fines and penalties for environmental violations. Although consensual methods for fulfilling public functions have gradually gained traction, command-and-control methods continue to hold a prominent role in preventing and penalizing legal violations. Environmental licensing and Environmental Impact Assessments were also thoroughly examined as essential regulatory mechanisms for managing activities that affect the environment. In Brazil, any activity that exploits environmental resources actually or potentially polluting or capable, in any way, of causing environmental degradation, must undergo prior environmental licensing. Environmental Impact Assessments are required only for activities likely to have significant environmental impacts. Environmental zoning was also discussed as an important regulatory method aimed at rationalizing land use and enhancing environmental protection. Finally, economic tools for environmental regulation were analyzed as strategies to encourage sustainable and environmentally protective behaviors or to deter unsustainable and environmentally harmful practices, thereby fostering voluntary cooperation from private individuals, albeit through incentives. Particular attention was given to environmental taxation and payment for environmental services.

Chapter 9 delved into the important topic of environmental civil liability. This concept corresponds to *responsabilité civile* in civil law systems or tort law in common law systems, examining criteria for determining when damage caused to others should be remedied and how it should be addressed. Compensation for environmental damage in Brazil is typically pursued through class actions, particularly via the *public civil suit*. Readers learned that in Brazil, the scope of compensation for environmental harm is broad and applies irrespective of fault or intent, with reparations achievable through various means, both pecuniary and non-pecuniary.

Chapter 10 addressed criminal environmental liability. Many environmental offenses are classified as crimes and can be penalized accordingly, although most carry lighter penalties, allowing for *soft measures* such as alternative sentencing and criminal agreements. A distinctive feature of Brazilian law is the possibility of holding companies or legal entities criminally liable, though only for environmental crimes.

Judges and courts play a crucial role in shaping Brazilian environmental law. Chapter 11 highlighted that the constitutional framework allows for extensive judicial scrutiny over public administration actions and inactions, including environmental matters. Beyond the right of access to justice, the Constitution enshrines the separation of powers and judicial independence, and grants substantial protections to judges. Broad access to justice in environmental matters is enabled by multiple legal actions, such as public civil suits, people's legal suits, judicial review, and more. Consequently, the study of *Environmental Courts & Tribunals* finds fertile ground in the Brazilian judicial system.

Chapter 12 provides an in-depth examination of specially protected areas, highlighting deforestation and land-use change as key contributors to greenhouse gas emissions in Brazil. Brazil, with the Amazon—the world's largest tropical rainforest—holds unparalleled biodiversity, emphasizing the significance of its protected area legislation. Consequently, this book pays particular attention to statutes like the Nature Conservation Units Act (Law n. 9.985/00) and the Forest Protection Code (Law n. 12.651/12), which serve as the primary legal framework on the topic. The 1988 Constitution was pivotal in mandating that the government designate specific territories and components for special protection. It further stipulates that alterations or reductions in these areas are permissible solely through legislation, prohibiting any use that could compromise the attributes that warranted their protection (1988 Constitution, Art. 225, § 1, III). This constitutional directive significantly boosted the enactment of conservation laws and the expansion of environmental protection zones across Brazil.

Chapter 13 examined climate change law and climate litigation. Following the trajectory of international law, Brazilian legislators and courts are increasingly addressing climate change. Brazil has participated in all major international climate agreements, from the 1992 Framework Convention on Climate Change (UNFCCC) and the 1997 Kyoto Protocol to the 2015 Paris Agreement and subsequent conferences. While Brazil has a Climate Change Act (Law n. 12.195/09), it still lacks specific targets for public authorities and companies. Nonetheless, promising actions lie ahead, including the structuring of a carbon market, refining carbon taxation, and expanding renewable energy development. Moreover, climate litigation has considerable potential, further opened and encouraged by the Supreme Court's decision in the *Climate Fund case* (ADPF 708) and notable judgments aimed at safeguarding the Amazon Rainforest and the Pantanal from deforestation and wildfires.

Finally, Chap. 14 offers insights on the regulation of water resource exploitation. Brazil plays a crucial role in this area, being one of the most water-abundant countries globally. The challenge lies in ensuring the rational and sustainable use of waterways, prioritizing essential uses while preserving the environment and safeguarding life.

Glossary

ADC. ***Ação Declaratória de Constitucionalidade*** It stands for a lawsuit filed directly before the Supreme Court, whereby a declaration of constitutionality of a law is sought. It is the mechanism to obtain from the Supreme Court a declaration of constitutionality of a law, to avoid its questioning, in the opposite way to an ADIn.

ADIn. ***Ação Direita de Inconstitucionalidade*** It stands for a lawsuit filed directly before the STF, whereby a declaration of unconstitutionality of a law is sought. It is the mechanism to obtain judicial review of the constitutionality of the law.

ADO. ***Ação Direta de Inconstitucionalidade por Omissão*** **(Action for Unconstitutionality by Omission)** It stands for a lawsuit filed directly before the STF, whereby a declaration of unconstitutionality on account of lack of a measure to render a constitutional provision effective is sought.

ADPF. ***Arguição de Descumprimento de Preceito Fundamental*** **(Claim of Non-compliance with a Fundamental Precept)** It represents a kind of lawsuit before the Supreme Court to complain of state action or omission violating a fundamental constitutional right.

AgInt no AREsp. ***Agravo interno no agravo em recurso especial*** It stands for a kind of appeal against a Superior Court of Justice decision to be decided for another body of judges within the Superior Court.

AgRg na SLS. ***Agravo Regimental na Suspensão de Liminar e de Sentença*** It is a kind of appeal against a Superior Court judge's decision in a juridical remedy called "*Suspensão de Liminar,*" which is adopted to suspend an interim relief. It is decided by a body of judges in the same Court.

ANVISA. ***Agência Nacional de Vigilância Sanitária*** **(National Health Inspection Agency)** It is a federal regulatory agency whose purpose is to supervise the production and consumption of products subject to sanitary regulation, such as drugs, agrochemicals, and cosmetics, as well as to perform sanitary control at ports, airports, and national borders. It is regulated by Law n. 8.080/1990.

Art. ***Artigo*** **(Article)** It stands for a subsection of a statute or a legal act.

R. Martins Costa Moreira, G. Wedy, *Brazilian Environmental and Climate Change Law*, https://doi.org/10.1007/978-3-031-87998-2

CONAMA. ***Conselho Nacional do Meio Ambiente*** **(National Council for the Environment)** It is a consultative and deliberative body, created to advise, study, and propose to the Government Council, government policy guidelines for the environment and natural resources, and promulgate, within its jurisdiction, environmental regulations, and standards.

CPC Code of Civil Procedure (Law n. 13.105/2015).

CRA. ***Cota de Reserva Ambiental*** **(Environmental Reserve Quota)** It is provided for in the Forest Protection Code (Law n. 12.651/12) and basically allows compensation for the exploitation of a rural property with suppression of the *Rural Property Preservation Area* (called *Reserva Legal* by the statute) to an extent above that allowed by law, by preserving this protected area on another property, to an extent greater than legally required.

Dec. ***Decreto*** **(Decree or Order)** It is a kind of administrative act promulgated by the President of the Republic. In the past, before the current Constitution, the *decretos* had the same force as a statute passed by the National Congress. This is no longer the case under the current Constitution. Currently, the *decreto* is inferior to the Act of Parliament and cannot innovate or contradict the legislation (*lei*).

IBAMA. ***Instituto Brasileiro do Meio Ambiente e dos Recursos Naturais Renováveis*** **(Brazilian Institute of Environment and Renewable Natural Resources)** Created by Law n. 7.735/1989, IBAMA is a federal agency linked to the Ministry of the Environment, with the function of exercising supervision and environmental protection, to execute actions of national environmental policies, referring to federal functions, related to environmental licensing, environmental quality control, authorization for the use of natural resources, and environmental supervision, monitoring and control, observing the guidelines issued by the Ministry of the Environment.

J. Judgement This indicates the date of the ruling.

Inc. ***Inciso*** It indicates a subdivision of an article or a paragraph of a statute or a legal act.

LC. ***Lei Complementar*** **(Supplementary Law)** *Supplementary Laws* differ from other *Ordinary Laws* (*Leis Ordinárias*) in that they require the vote of a majority of the members of the House of Representatives and the Federal Senate to be approved. They must be adopted to regulate specific subjects when expressly determined in the Constitution of the Republic. It is provided for in Art. 69 of the Constitution.

Lei The same as a *legal act* or *statute* in US law; it is the law that is passed by the National Congress.

Min. ***Ministro*** **(Minister)** It designates the judges of the higher courts (e.g., Supreme Court and Superior Court of Justice). Similar to the *Justice* in US law.

Par. ún. ***Parágrafo único*** **(sole paragraph)** It indicates the only subdivision of an article of a statute or a legal act.

RE. ***Recurso Extraordinário*** It is a kind of appeal against an Appellate Court, or a Superior Court decision filed before the Supreme Court, to decide on controversies involving the interpretation of the Brazilian Constitution.

Res. ***Resolução*** **(Resolution)** It stands for a kind of regulation as a form of a rule, promulgated by administrative branches or councils, like CONAMA.

REsp. ***Recurso Especial*** It is a kind of appeal to be filled against an Appellate Court decision before the Superior Court of Justice, to decide on controversies involving the interpretation of the federal law (not including constitutional issues, which are the object of a *Recurso Extraordinário* before the Supreme Court).

STF. ***Supremo Tribunal Federal*** The Supreme Court of Brazil, based in the capital of the country. It is assigned jurisdiction for constitutional issues, among other functions, pursuant to Art. 102 of the Constitution. Its judges are called *Ministros*.

STJ. ***Superior Tribunal de Justiça*** **(Superior Court of Justice)** The STJ is a superior court, based in the capital of Brazil, regulated in Art. 105 of the Constitution, with jurisdiction, primarily, to standardize the interpretation of federal legislation by means of an appeal, among other attributions. Its judges are called *Ministros*.

The manufacturer's authorised representative in the EU is Springer Nature Customer Service Centre GmbH, Europaplatz 3, 69115 Heidelberg, Germany. If you have any concerns regarding our products, please contact ProductSafety@springernature.com

Printed and bound by CPI Group (UK) Ltd, Croydon, CR0 4YY
30/09/2026
02225558-0004